Shiva's Hologram

The Maheshwara Sutra

Shiva's Hologram

The Maheshwara Sutra

Padma Aon Prakasha

BOOKS

Winchester, UK
Washington, USA

JOHN HUNT PUBLISHING

First published by O-Books, 2023
O-Books is an imprint of John Hunt Publishing Ltd., 3 East St., Alresford,
Hampshire SO24 9EE, UK
office@jhpbooks.com
www.johnhuntpublishing.com
www.o-books.com

For distributor details and how to order please visit the 'Ordering' section on our website.

Text copyright: Padma Aon Prakasha 2022

ISBN: 978 1 80341 334 1
978 1 80341 335 8 (ebook)
Library of Congress Control Number: 2022942582

A CIP catalogue record for this book is available from the British Library.

Design: Lapiz Digital Services
Cover Image Copyright Padma Aon Prakasha
Cover Design and Creation by Aina Greta

UK: Printed and bound by CPI Group (UK) Ltd, Croydon, CR0 4YY
Printed in North America by CPI GPS partners

The author of this book does not dispense medical advice or
prescribe the use of any technique as a form of treatment for
physical, emotional, or medical problems without the advice of a
physician, either directly or indirectly. The intent of the author
is only to offer information of a general nature to help you in
your quest for emotional and spiritual well-being. In the event
you use any of the information in this book for yourself, which is
your constitutional right, the author and the publisher assume no
responsibility for your actions.

We operate a distinctive and ethical publishing philosophy in
all areas of our business, from our global network of authors to
production and worldwide distribution.

Contents

The Dance of Shiva

It is said...

A long time ago, Shiva called forth the greatest sages of India, desiring to elevate them into the divine matrix of creation through the rhythmic intelligence of the Tandava Dance of destruction and creation.

He smiled at the assembled Avatars and Rishis, who watched on in awe. Moving his limbs gracefully, slowly, he assumed the gracious yet terrible Nataraja form and posture, hands held aloft, limbs in perfect equipoise.

Slowly at first, his arms started to unfold, weaving patterns of unutterable beauty and awesome power, forming the mudras of the *Tandava*.

Gaining speed with each successive gyrating pose, each whirl of the arms, each spin around his axis, the air started to crackle and burn, fizzing electric intensity, revealing fleeting glimpses of space and the stars as he whirled. Space itself began to be rent asunder, revealing the looming vastness of the void.

Emblazoned in an iridescent cocoon of fire, radiating light, whirling faster and faster, beyond the speed of light, beyond space, time and our universe's laws, Nataraja-Raja span so fast he became still.

In this Singular moment, his wild movements were caught in freeze frame. He was Dancing on the broken, limp body of the prostrate dwarf demon *Apasmara*, whose face was contorted into a grimace of crazed glee, eyes raw, red and bleeding. He was the embodiment of forgetfulness and fear, amnesia incarnate.

Nataraja-Raja gracefully arced his foot up high, thunderously smashing it down into Apasmara, crushing him into myriads of atoms of incandescent light. The vast expanse of space, of consciousness itself, became illuminated by this fire of all fires, the fire that burns the karma of those beyond karma, in a single, infinite moment.

Everything stopped in the Thunder of Silence.

In this moment of complete and utter stillness, Shiva Maheshwara held out his upraised right hand and sounded nine beats on his hourglass *Damaru* Drum. Spiralling out from the Silence, these pulse waves of sonic consciousness uncurled and unfolded into luminescent shapes and letters of light weaving exquisite matrices.

He sounded five more beats, thus composing 14 beats in perfect rhythm, echoing throughout the vastness.

This Blueprint of creation are the sounds of the Maheshwara Sutra.

The Dance of Shiva is the dancing universe, the ceaseless flow of energy going through an infinite variety of patterns that melt into one another.
– Fritjof Capra

In the Dance, Nataraja-Raja arises from the void surrounded by the flaming sphere of light *prabhu mandala*, the unbroken continuum of creation and destruction. This *prabhu mandala* is a portal into Aum, the sound current field of physical creation.

In the Dance, "every subatomic particle does an energy dance *and is* an energy dance; a pulsating process of creation and destruction without end... a rhythm of creation and destruction manifest in the birth and death of all living creatures, and the dynamics of subatomic particles."[1]

Nothing in creation is ever destroyed. No energy is ever lost; it just changes its expression in an alchemical dance from one movement to another. Within our own evolution, parts of who we are have to "die" and be transformed in order for new aspects of us to bloom. The dance of destruction is therefore more appropriately called the *dance of transformation* from one thing, energy or process, into another.

Creating and destroying are two poles of a single unified cycle. Creating and destroying mark out the beginning, and then the completing of a sphere, composed of an inflowing wave that creates, and an outflowing wave that dissolves.

From within Shiva or pure awareness arises creation, its objects, contents and experiences, which then dissolves back into awareness. Creating and dissolving are two sides of the same coin, each one being portals or wormholes into the universal process: doorways into the infinite.

Shiva's Dance shows these dynamic processes constantly transforming everything all the time, such as the stars, in nature, and in our bodies, where our cells die, renew and transform every day. We are dying and being reborn every day, transforming every day, even right now whilst you are reading this! To harness these forces is one of the goals of science, and to be able to refine, use and transcend the pull of these forces is the goal of all enlightenment traditions.

The Dance of Shiva has two parts to it. The first and most well-known is the *Tandava* Dance, where Shiva dances and destroys the macro universe *and* the micro universe of the human being by dancing on the demon dwarf *Apasmara*, the embodiment of all our illusions, conditionings, karmas and ignorance.

Shiva has his foot firmly on Apasmara's body as he dances, with Apasmara's contorted face, gaping mouth and red eyes looking up at him in a somatically demented, epileptically insane fit of anger, madness and chaos.

In India, it is considered a great blessing to be under the foot of Shiva, to be annihilated by Him into cosmic consciousness. This blessing was an act of total surrender and what devotees[2] of Shiva aspired to: to totally give themselves to Shiva so their separate self could be destroyed in his Dance, as *"He destroys all forms of karma in His Dance being seen."*[3]

Yet Shiva's Dance, so iconic and well known across the world, would not be so popular if people knew what it actually does to you if you experience it! Indeed, South Indian accounts of those who have witnessed the Dance range from abject terror to ecstasy.

Priests, kings, sages and Rishis would involuntarily fall to the ground, trembling and shaking, entranced and lost to the world. Others would become totally fear stricken, catatonic, limp and broken, like Apasmara. To experience the Dance of Nataraja-Raja is to have your self, and your world, totally transformed. The Dance is the process and the fuel of all alchemy. By surrendering to the Dance, your consciousness and identity dissolves.

The Maheshwara Sutra tells the story of what comes after this Dance, when he bangs out on his Drum the sounds of the new creation.

The Audience

The importance of the Maheshwara Sutra can be gauged by the audience present at its revealing, the cream of Vedic yogis, avatars and Rishis. Shiva called forth the greatest hearts and minds of Vedic India to witness his Dance, desiring to elevate these perfected beings or *Siddhas*, who had already mastered many of the spiritual powers that are part of the highest human flowering and potential.

The wisdom and workings of the quantum blueprint of creation that unfold throughout the Maheshwara Sutra are found in the teachings and practices of the sages, Rishis and Avatars who were present to witness destruction in the Tandava Dance, and the subsequent 14 beats of Shiva's Drum that issued forth the new creation in the Maheshwara Sutra.

Bringing their non-dual wisdom together with the direct experience of these 14 waves of creation allows the story of creation, the story of the Maheshwara Sutra, to be told.

The first and most well-known of those mentioned in the prologue to the Maheshwara Sutra is *Patanjali*, the writer of the *Yoga Sutras*, the classic and essential text on Raja Yoga and Self Realization. He is also the author of the *Caraka Samhita*, the root text on *Ayurveda*, the science of life, health and healing.

Patanjali also penned a masterwork on Sanskrit Grammar inspired by the Maheshwara Sutra called *Maha Bhashya*, which simplified the programming code of Sanskrit Grammar. He is mentioned in the Maheshwara Sutra as much of his wisdom explains the Sutra and derives from the Sutra as well.

Patanjali is intimately connected to *Chidambaram*, the "palace of sky consciousness and cosmic intelligence", a Shiva Temple in South India where the Tandava Dance occurred. Patanjali wrote the *Patanjalipuja Sutra*, the priests' manual for daily worship at Chidambaram Mandir, and this ritual manual is still used in ceremonies and festivals today, being called the *Citsabhesvarotsava Sutra.*

Patanjali's title in ancient India was "Lord of the Serpents", for he was a manifestation of the Serpent Power like Kundalini. Serpents represent powerful forces of consciousness and shakti. As symbols of transformation like the Dance of Shiva, they shed their skins, what no longer serves them, in order to transform into something else. No energy is ever lost – it just transforms.

Serpents have to shed their skin to transform, otherwise they die. It is part of their innate make-up, and a reminder to us all of our own journey. Transforming what does not serve us allows us to thrive, bloom and grow. Having serpents wrapped around one's body indicates one is in this process, this flow of creation and dissolution that defines the Dance and nature.

Patanjali was the human incarnation of Vishnu's supporting serpent *Shesha,* the serpent that Vishnu rests on in the "milky ocean" of the quantum field before the creation of a new physical universe. As Ananta Shesha, "Endless-Shesha", or Adishesha, the "First Shesha" uncoils, time moves forward and creation takes place; when he coils back, the universe ceases to exist.

As the story goes, when Vishnu told Shesha his story of seeing the Tandava Dance in the Daruvana forest, Shesha was so moved and inspired that he asked Vishnu's permission to go and see the Dance for himself. After being granted permission, Shesha went into deep prayer meditation to Shiva, who then gave Shesha the boon of being born to the great Rishi Atri and his compassionate wife *Anasuya.*

When Patanjali was born, she dropped him on the ground in shock as he appeared in his half-snake form in her hands, from where comes his name Pata-anjali, "fallen from folded hands".

When he was old enough, Patanjali/Shesha made his way to the sacred forest of *Thillai-Chidambaram* (the place where the Dance was set to happen) through *Patala*, the underworld of the serpents or *nagas*. When he reached Thillai-Chidambaram he was met by the Sage *Vyaghrapada*, the Tiger Footed One, son of the Sage *Madhyandina*.

Vyaghrapada is mentioned in the orally passed down prologue to the Sutra as a great devotee or *bhakta* of Shiva present for the Dance. Vyaghrapada acted as Patanjali's guide when he first arrived to the Thillai forest, orientating him to the sacred surroundings.

Vyaghrapada and Patanjali were a duo of Masters, doing ceremonies together, meditating together and building lingams and other structures in Thillai-Chidambaram whilst waiting for Shiva to Dance. Perhaps they prepared a way for the Dance to occur by creating a consecrated sacred space and environment for it to happen. It is only in response to sincerity, devotion and dedication that deep change can happen.

BrahmaRishi *Vasistha* is also mentioned in the oral prologue to the Maheshwara Sutra as being present at Shiva's Dance and subsequent re-creation. Vasistha is one of the seven main Rishis of Vedic India, a Mind Born Son of Brahma, born directly from the mind of Brahma, not born from a human womb.

Rishi Vasistha was created to continue Brahma's role, which included propagating life on earth, educating humanity, *and* to produce more children (over 100 of them!). Vasistha was created *in order to create*.

Vasistha was the composer of some of the *Rk Veda* Hymns of Creation, and the Guru of Patanjali, Lord Rama and the entire *Suryavansha* Solar Dynasty. His role as teacher and High Priest of the Suryavansha Solar Dynasty enabled an unbroken succession of pristine wisdom from the beginning of creation to be passed down from generation to generation.

Rishi Vasistha is one of the most important Rishis in *Saivite Dharma*. The Seer of one of the most famous of Shiva's mantras, the *Tryambakam*, Vasistha and his followers were known as

Kaparda, wearing matted locks on top of their heads, just like Shiva. His book *Yoga Vasistha*, which records his teachings to Lord Rama, is one of the pillars of Vedic wisdom.

Rishi Vasistha is the Rishi with the deepest connection to both Shiva and Brahma in their roles of creator and destroyer, and he understood and practised these himself, teaching aspects of this wisdom to his students. As the Guru of Patanjali he helped shape him into the Master he is.

Rishi Vasistha is mentioned in the prologue to the Sutra for all these reasons, and because some of the wisdom he holds derives from the Sutra *and* some of his wisdom explains the Sutra as well.

The revelation of the blueprint of creation was also given to *Sanaka Kumara* (specifically mentioned in the first verse of the Sutra), one of the Four Kumaras or Avatars of Brahma, born directly from the mind essence of Brahma.

These four were created directly from Brahma's consciousness, not from a human womb, to continue Brahma's work of propagating creation and giving humanity a pathway into God consciousness.

Brahma immaculately conceived and directly begot the Kumaras as *saktyavesa avataras*, or indirect incarnations of God. When God comes directly to earth in form, as with Krishna and Rama, He is called an avatar or *saksat*, and when God empowers a living entity to represent him as an agent of sorts, such a being is called an indirect or *avesa* avatar, beings who are invested with the power and Law-making abilities of God.

Once they had been created, the Kumaras looked at humanity and the world and were not interested in joining in with what was happening. They did not want to have families, offspring or sex. Yet, they also saw the nature of human suffering, understanding that their reason for being created was to help humans find ways into their source and God consciousness.

As avesa avatars of our planet Earth and humanity, the Kumaras periodically bring us new teachings and embodiments of the divine in order to help guide, inform and elevate us into becoming a conscious part of God Consciousness. They have been doing this for many thousands of years.

As sublime embodiments of God and teachers of awakening, the Kumaras have no equal. In the *Srimad-Bhagavatam* 2:7:5 it is said that, "these four are incarnations of the knowledge of the Supreme Lord, and explained transcendental knowledge so explicitly that all the sages could assimilate this knowledge without difficulty. By following in the footsteps of the four Kumaras, one can see the Supreme Personality of Godhead within oneself."

Kumara means "child" and they are always pictured in Indian literature as looking like five-year-old children, beatific, innocent and full of love, like cherubim angels.[4] They are seen, along with Krishna and Visnu, as being pure love embodiments of the divine.

As it is said, Brahma created the four Kumaras to propagate and help humanity by empowering them to share four streams of wisdom: the science and study of creation; yogic mysticism for liberation of the soul; the art of detachment and witnessing, and meditational practices in order to attain enlightenment.

The Kumaras then inaugurated their own spiritual lineage, the *Kumara-sampradaya* or *Nimbaraka-sampradaya*,[5] and spread these teachings in order to share pathways into the source of creation and God Consciousness.

For example, Sanat Kumara revealed the direct and fast path to enlightenment in Kashmir Saivism through his transmission of teachings to Rishi Durvasa, as Swami Lakshmanjoo shares. Sanandana Kumara revealed and taught the original mystic Christianity in his incarnation as Christ Yeshua, as the

Theosophists share. The Kumaras have even more hidden significances, as we shall discover later in this book.

Another illustrious personage mentioned in the Sutra is the Sage *Panini*, who developed the code or rules of Sanskrit Grammar after being present at the Dance and the subsequent Creation detailed in the Maheshwara Sutra. It is through sounds, with their infrastructure code of grammar, that the universal matrix and human matrix are created, and Panini was a master of this code.

The final person mentioned in the Sutra is *Rishi Nandikēśvara*, who wrote the *Nandikēśvara Kāsikā*, the original commentary on the Maheshwara Sutra. His commentary was written after the sages who had received the Maheshwara Sutra approached him to understand the meanings of the 14 classes of sound. They all considered Nandikeshvara to be the right being to convey these secrets revealed by Shiva.

In response to the sages, Nandikeshvara or *Nandi-natha*, named after the Nandi Bull of Shiva who guards the entrance to his mandirs or temples, wrote the 27 verses of the *Nandikeshvara Kasika*, adding the meanings to each sound in the Sutra.[6]

Rishi Nandikeshvara lived around 250 BC, and is one of the first Gurus of the *Saiva Siddhanta* tradition. He is recorded in Panini's book of grammar as being a Guru of two of the greatest Vedic Gurus in Patanjali and Vasishtha, who passed down much of the knowledge of Self Realization to us today.

Nandikeshvara is referred to as the "rhythm master" of Shiva. He is frequently pictured with a drum by his side, and even today in South India is honoured as one of the principal Gurus in music concerts. He systemized classical Indian Dance as an expression of the moods, flows and movements of universal energies in his text *Mirror Gestures*, which today is one of the foundations of Indian dance and dramatic techniques.

This Commentary

The original commentary or *vārtika* on the Maheshwara Sutra was written by Rishi Nandikeshvara as a response to the questions posed by the sages present at the Dance of Shiva. This Commentary became known as the *Sri Nandikesakasika*.

The sage Upamanyu then wrote a vārtika commentary in the eleventh century on the Maheshwara Sutra called *Tattva Vimarśinī*. In 1999, a third vārtika named *The Sounds of Shiva* was written by Peter Harrison, a London-based Sanskrit scholar, author and teacher.

I was introduced to the Maheshwara Sutra by my friend and colleague Peter Harrison, author of *Dhatu Patha* with Simon Hill. He was the owner of Golden Square Books, London's premiere Vedantic bookstore, and later went onto become a manager of Watkins Books, Europe's most eminent metaphysical bookstore and publisher.

The first time we met, there was an instant knowing and connection between us. We quickly became friends and enjoyed many lunches at Peter's home with his family. It was Peter who first mentioned the Maheshwara Sutra to me, and one day he casually gave me a copy of his own commentary on it.

As I read the first two verses of the Maheshwara Sutra, my whole body and soul lit up. I spontaneously started to tremble in excitement and glee. Within seconds, every cell in me knew it was part of my Dharma to write about this.

I Am grateful to Peter Ji, who gave me his blessing to write this book after I shared my feelings about it to him. His self-published *Sounds of Shiva* was a reference point for this book, as was the *Tattva Vimarsini* of Upamanyu, and of course Rishi Nandikeshvara's original *Nandikakasika* commentary.

*

All three of these vārtikas are referenced in this fourth vārtika: *Shiva's Hologram*. These four are the only available vārtikas on the Maheshwara Sutra that have been made available to the public, as the Sutra has been kept within Saivite Initiatic circles in India, known only to a few, until now.

There are two Sanskrit versions of the Sutra, with minimal differences between the two; I have incorporated both. There is also a strong grammar aspect to the Maheshwara Sutra that the founder of Sanskrit grammar *Panini* expounded upon in his treatise *Astadhyadi*, which I leave to the reader to research if they are so interested.

The structure of this vārtika *Shiva's Hologram* is slightly different in verses 3-9, as I have grouped together the verses expressing the first sounds of creation *AIUn* in order to clearly explain each sound and its role in creation. Other than that, this vārtika follows the narrative flow of the original *Sri Nandikesakasika* by Rishi Nandikeshvara.

The purpose of this vārtika *Shiva's Hologram* is to bring this timeless wisdom into the quantum understanding of the 21st century, making its wisdom contemporary and understandable without diluting the essence of the Sutra in any way.

It can help develop our understanding of the quantum universe and the true nature of reality, and can help us build a quantum civilisation in virtually all fields of endeavour, from technological to spiritual, from linguistics and advanced communication abilities to sonic consciousness, health, healing, mindfulness and yoga.

This commentary is not religious, but rather part of the non-dual *Saivite* lineage that the Maheshwara Sutra is based on, and that my ancestors and I are part of as well. The head of my *gotra* or bloodline is Rishi Vasistha, one of the Vedic Seers present at the revealing of the Maheshwara Sutra. His wisdom is a key to the Maheshwara Sutra, and his loving presence with me as my great father has been a solid lineage support for this vārtika and myself. I Am forever grateful to him for his love and help.

The Maheshwara Sutra is virtually infinite. One could write and develop a commentary and teaching on this Sutra for one's whole life, and still not cover all the aspects that the Sutra describes. I apologize in advance to anyone who feels aspects are left out of this commentary, but no one human could write a complete Theory of Everything, or a complete Story of Creation, which is what the Maheshwara Sutra shares.

It is said in *Saivite* lore that the Maheshwara Sutra contains all the sciences and secrets of creation within it, and each person will understand it through their own unique lens of perception. To some people, the Sutra reveals what quantum physics is now beginning to understand. To others, it reveals the unity of creation in *advaitic Saivism* or non-dual unity consciousness.

To others, it reveals the elegant perfection of the Sanskrit language and how sound creates us and the universe, and to others it reveals the systems behind reality. It all depends on your inclination and state of consciousness, the waveband of frequency you live in, as to how you will understand the wisdom of the Maheshwara Sutra.

For yogis, the Maheshwara Sutra is experiential, a grand tool and map to access the sonic consciousness of creation through the direct experience of the sounds, mantras, meanings and wisdom of the Sutra. This map can help elevate oneself into the AHAM, the I AM or soul *Atma*, and then beyond this state into Creator or God Consciousness. This is the greatest use of the Sutra, and where its true power lies.

The Drum

नृत्तावसाने नटराजराजो
ननाद ढक्कां नवपञ्चवारम्।
उद्धर्तुकामस्सनकादसिद्धा -
नेतद्विमर्शे शिवसूत्रजालम् ॥ १॥

Nrttavasane natarajarajo nanadadhakkam navapancakaram
Uddhartukamassanakaisiddha netadvimarse sivasutrajalam

1

After his Dance, *Nataraja Raja*, King and Lord of the Dance,
beat his drum nine and five times
to elevate *Sanaka* and other perfected ones
into the fluid matrix of Siva's Sutras.

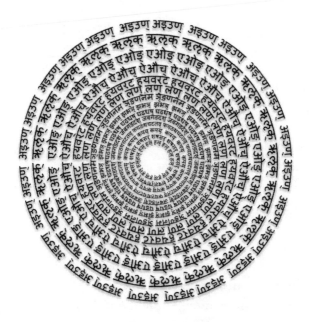

The nine and five sounds from Shiva's *Damaru* Drum emanate forth the sonic blueprint for creation and the human being, resounding these sound vibrations into being. These sounds spring forth and weave this web of vibration that arises from Nada-Bindu,[7] the Singularity point and origin of creation.

Shiva's *Damaru* Drum is an hourglass shape. The hourglass represents the flow of time, and the drum measures the processes of time, which evolve and dissolve forms. Space holds the container for time within the hourglass shape, and space is the medium for energy to form matter through the processes of time.

Space is as important to music as are notes and sounds, as any composer knows, for the space between sounds, *its timing*, forms the shapes of music. Today, scientists have concluded that space is at the heart of all objects, and we are, in essence, over 94% empty space, with a covering of matter on the surface. This "dark matter" is that which is invisible to the senses and the mind.

In the hourglass shape the upper chamber represents inward, invisible vibrations, whilst the lower chamber represents matter, visible vibrations that are outward orientated. The passage through which the sands run in the hourglass connects both chambers and is their point of meeting, where the inward becomes outward and visible, and the outward and visible becomes inward and invisible.[8]

The hourglass is the meeting of two: one downward facing, one upward facing, one masculine and one feminine: the dual nature of life, *and* what is required to bring forth life. One is the erect masculine *phallic nature*, the other the feminine womb, a fertile incubatory space to gestate and grow.

These two polarities also resemble two toroidal spirals, expanding and contracting, breathing in and breathing out: the creative process of the universe and our heartbeats. Within the hourglass, all natural processes and cycles occur.

Bindu

NavaYoni Yantra
9 Cosmic Wombs

Shiva's Drum
Rhythm of Creation

The passage through which the sands run that connects both chambers is where the masculine meets the feminine, where human meets the divine, where life and death meet simultaneously, and where creation begins and ends. This Dance is of masculine and feminine, Shiva and Shakti, creating the universe, for with both in union, all creation reveals. The hourglass drum contains all creation *within* it, and from it comes forth all creation.

In Indian lore, the damaru is a channel for the blessings and power of the *Saivite* lineage and the forces that protect this dharma pathway. In Southern Indian texts, the celestial musician or *Gandharva Citrasena* learned the entire science of music from hearing the damaru. Interestingly, when one physically plays the damaru, it produces dissimilar sounds which through resonance create one sound. This sound symbolises Nada, the cosmic sound.

It is also said in ancient Tamil scriptures that Shiva is the constellation Orion. Orion is the constellation where many stars are produced – a centre of creation. The shape of Orion looks remarkably similar to Shiva's *Damaru* Drum. Coincidence? We will find out more about this at the end of the book, where it will make more sense.

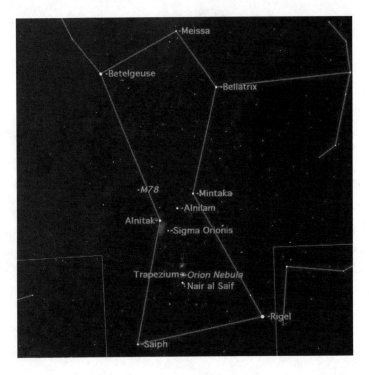

In the silence after the Dance of dissolution, Shiva resurrects and upgrades creation and the human form by weaving a web of sound vibrations through the Drum to create a divinely ordered,[9] harmonious matrix pattern for creation in[10] a living mesh.

This network database formed by the Supreme network engineer Maheshwara is a web, network and grid matrix of vibration that springs forth like a fountain (*jalam* means fluid and water).[11] This matrix of vibrations is formed by seed-syllables or *bijas*. Seeds hold the blueprint for evolution in a latent form, and one meaning of the word "evolution" refers to an unfolding of a seed form.

The 42 seed syllables of the Maheshwara Sutra are named and threaded together throughout the Sutra to form a complete understanding of sound consciousness and how vibration creates life. Each seed holds a specific process, function and power of creation within itself, and each seed holds a form of consciousness in a latent design until it is activated and expanded to flower into waves of manifestation, expanding from these seed designs to manifest these designs in all dimensions.

These 42 *bijas* generate the shapes and structures of consciousness and matter through waves of vibration. These shapes of consciousness are the holographic matrix of creation, and were seen millennia ago in Indra's celestial Net:

This imperial net is made of jewels. Because the jewels are clear, they reflect each other's images, appearing in each other's reflections upon reflections, ad infinitum, all appearing at once in one jewel, and in each one it is so – ultimately there is no coming and going. It is precisely by not leaving this one jewel that you can enter all the jewels. If you left this one jewel to enter all the jewels, you couldn't enter all the jewels. Why? Because outside this jewel there are no separate jewels... if you take away this jewel there will be no net.

Each jewel is a node point, a seed syllable that connects the holographic web throughout space and time. *Bijas* are concentrated essence points from where consciousness arises from, and from where the entire universe is grown, flourishes, and is revealed in all its nuances and levels of meaning and application.

This idea is echoed in the human brain, a complex interacting network that relies on nodes to convey information from place to place. Very few "jumps" are necessary to connect any two nodes, as researchers discovered using functional magnetic resonance imaging (fMRI) to study how different brain regions connect, and the degree of correlation between activities in tens of thousands of brain regions.

They found that a small number of nodes were connected to many others. These "super-connected" nodes act as hubs – as with the Internet – getting the "word" out quickly and widely.[12] "This so-called 'small world' property allows for the most efficient connectivity," said Dante Chialvo, a physiologist at Northwestern University. Other networks – social and biochemical – rely on the same principle.

In the Maheshwara Sutra, syllables are the basis of all forms, thoughts, perceptions and expressions, collectively weaving together the underlying meaning of who we are and the constructs of the matrix of creation.

We bring these vibrations into the third dimension through sound syllables and letters, which create words. Words create thoughts, thoughts create language, and language shapes, informs and guides our perceptions of the world we live in. We cannot think or express without words, and word vibrations need an alphabet, a structure, a matrix, to bring it into form and make it understandable.

Sanskrit: The Alphabet of Creation

The Alphabet of Creation revealed in the Maheshwara Sutra shows how the universe and ourselves are dynamically formed, shaped and created through sound vibration.

The 14 classes of sound that contain the 42 *bijas* or seed syllables in the Sutra describe all the ways the human voice can mould sound vibration into expression, *and* show how vibration weaves the cosmos and the individual together. These 42 sounds give form and boundaries to all the different flavours and movements of consciousness across the whole gamut of human expression.

Throughout history, we have expressed and tried to understand these movements of consciousness through sound, linguistics, metaphysics, religion, music, geometry, sexuality, quantum physics and neuroscience. In the Maheshwara Sutra, all these fields of inquiry are underpinned by a spiritual science of consciousness formulated by the greatest minds and hearts of Vedic India through the perfected, elegant "language of the gods" known as Sanskrit.

Sanskrit is the basis of India's spiritual teachings, yoga, astrology, health and healing. It is the oldest intact language in the world, with the most literature ascribed to it, and contains many roots of the Indo-European language family, meaning that many words in English, Latin and Greek can be traced back to Sanskrit.[13]

The Maheshwara Sutra is the root code and foundation of Sanskrit, and is said to be one of the *Vedangas* or "limbs" of the Vedas, the supporting sciences behind the Vedas which unfold and explain the Vedas. All Sanskrit derives from this Sutra in its most precise form, showing us a wave-based language coursing throughout creation, unfolding the processes of matter and mind in a living, ecstatic, rhythmically intelligent matrix.

The Maheshwara Sutra paints a picture of the harmonious unfolding of vibration and meaning through the myriad forms and actions of the universal creative process, from the quantum field and beginning of creation to the forming of our three-dimensional reality.

Creation is Sound, and the Maheshwara Sutra gives us an entry point into this through the science of sound or *"siksa"*. This is a profoundly realigning process as the sounds of the Sutra compose the natural order of life itself.

Many scientists and spiritual leaders have called for a marriage of science and spirituality to solve the predicaments humanity finds itself in. The most scientific viewpoints taken by spiritual traditions the world over are found in Buddhism and Vedic culture, both of which share the language of Sanskrit to describe the quantum and invisible worlds.

As a vibratory language, Sanskrit is the most powerful, precise and practical means of activating the body-mind-spirit connection. It is not just a language as we think language to be. Sanskrit is a conduit between thought and the soul, between physics and metaphysics, between the subtle and material worlds, between nature and its creator.

Sanskrit is an experiential, fluid and felt language, bringing one into the direct experience of what is being communicated. Research has shown that the phonetics of Sanskrit have roots in the energy points of the body, and chanting Sanskrit stimulates these points, raising one's frequency and energy levels. This leads to an expansion of one's consciousness, enhanced brain functioning and energy levels,[14] improved health, memory, cognitive functions and reduction of stress.

As a holographic language Sanskrit is *apausreya*, or self-evident: it came before humans and stands independent of us, as the basis for the sounds of creation itself. As a holographic language, it is formed out of mantras, which through the power of our voices we can emulate, replicating the Sounds of Creation for ourselves.

The most important quality of Sanskrit is its effect on our state of consciousness, acting as a "tuning rod" for us to enter expanded states of awareness, mindfulness, peace and well-being. Speaking and listening to Sanskrit changes our consciousness, with every letter, syllable and word being an evocation of the fabric of creation.

For example, as Buddha sat under the Bodhi tree on the dawn of his enlightenment, he saw the Tree of Life stretch from Heaven to Earth, woven together by the glowing golden letters of the Sanskrit Alphabet.

When you immerse in the vibrations of Sanskrit, you are taken into the underlying structure of the matrix of creation; the effect can be like being "bombarded by thousands of prayers". The purity and harmonic resonance of Sanskrit show up as coherent waves on a sound spectrometer, and as exquisite, mathematically defined shapes as recorded in the science of Cymatics.

This means that the sound has a profound effect on our physiology and mind, as experienced by the hundreds of millions of people who use Sanskrit every day in order to enter expanded states of consciousness.

Even "people who are deaf have an exhilarating experience when they are exposed to Sanskrit chanting."[15] Even though they cannot hear it, they are experiencing the vibrations beyond audio and mental representations. Similarly, animals respond to Sanskrit chanting much more so than other languages. Try it and see what happens with your cat, for example.

Sanskrit sounds point to fluid, multidimensional qualities of consciousness rather than objects. Objects are the effect of qualities of consciousness. In other words, there is no separation between what a word sounds like, and what it means. This is yoga – when sound and meaning are the same. Thus, if you correctly sound a Sanskrit word, you will be taken into its pure consciousness at the same time.

Sanskrit gives a name to all things by the sound it actually makes, not by an artificial or arbitrary naming of an object. It names all things according to how they speak, how they sound, and how they transmit on multiple levels of vibration. If we know these sonic patterns, we can spend time in the vibrations that hold solutions to our problems.

Sanskrit is a multidimensional language with up to seven different layers of meaning for each word. These meanings will be interpreted according to the different degrees of consciousness each individual inhabits at that particular moment, i.e. one will understand and interpret each word from the level of their own consciousness.

Each sound is felt and experienced differently by the user when they are in differing states of consciousness. Thus, Sanskrit is a "travelling tool" into multiple states of consciousness and perception.

Mantra is the use of sound to manifest different states of consciousness. Specific mantras or "instruments to transcend the mind", like those found in the Maheshwara Sutra, are ever present, resounding around us and within us all the time – we just do not notice them, or are still enough to hear them.

All patterned sounds are mantras, and certain ever-present mantras can assist our consciousness into the world of universal patterns and quantum geometry, the invisible designs underlying the world of the senses and mind.

Certain mantras can lead the mind into this living light and ultimately silence, from where all sounds arise from, and dissolve into. We too can tune into these mantras all around and within us, and when we do, we enter the world of the Seer, poet and composer of the symphony of life, for the web of life itself is composed of these threads of sound.

Sanskrit is a sonorous and beautiful sounding language. In modern day India, pundits and priests recite it in prescribed metres and rhythms to achieve desired effects. This has now become the way that most people know Sanskrit to be pronounced: fast, monotonous, without beauty, sonority and a deeper consciousness. Most yoga teachers bark out Sanskrit in this way too, and entirely miss the vibrational coherence of Sanskrit.

Vedic Sanskrit is a sound vibration that comes from the depths of the Self: it is sonorous and emanates from the Self as a language of the Self. It does not have the same tonality as modern day Sanskrit rituals and does not have the same frequency and feel as English and other contemporary languages because Sanskrit does not arise from this mindset. Rather, Sanskrit shifts your attention into the Self when it is sounded properly and when it is received and allowed to be felt.[16]

Sanskrit has the highest number of vocabularies of any language in the world. Sanskrit is the most efficient and powerful language in the world, having the ability to express complexity concisely with the minimum number of words. Each word has up to seven layers of meaning and can therefore be translated in multiple ways depending on the level of consciousness of the translator; it often takes several paragraphs in English to explain a single Sanskrit word![17]

NASA has a department to research Sanskrit manuscripts because they believe Sanskrit to be the best computer friendly language (*Forbes* magazine, July 1987) because it is highly regularised. NASA declared Sanskrit to be the "only unambiguous spoken language on the planet" and the language most suitable for computer code and AI comprehension.

Eventually, whole algorithms could be based on Sanskrit when there is advanced technology to implement this through quantum computing.

Learning Sanskrit improves brain functioning and memory, as has been proved by Dr James Hartzell in his pioneering neuroscience studies. In practising Sanskrit, students start getting better marks in other subjects, as it enhances memory power. St James Junior School in London has made Sanskrit compulsory, and this has been followed by some schools in Ireland.

On a technical note, the introduction of new words into the Sanskrit language ended with the grammarian Panini, who laid down rules for the derivation of each word *and* stopped the introduction of new words by allowing for the joining (*sandhi-vicche*) of separate words to create entirely new wor(l)ds and meanings.

After Panini did this, some further changes occurred which were regularised by Vararuchi and Patanjali. However, any infringement of Panini's basic rules came to be regarded as a grammatical error and hence the modern Sanskrit language has remained the same without any change from the date of Patanjali (250 BC).

Sanskrit and the Maheshwara Sutra

There is One in whose hands is the net of Maya, who rules with his
power, who rules all the worlds with his power. He is the same at
the time of creation and the time of dissolution. Those who know
him attain immortality.

– Svetasvatara Upanishad

One of the beauties of Sanskrit is that each word can have
up to seven layers of meaning, and we can see this in the
multidimensional meanings of the word *Maheshwara* and the
word *Sutra*.

Maheshwara is one of the 108 Names of Shiva meaning the
"Great Lord, divine being, God". *Maha* means great, and *Ishwara*
means the Ruler, Lord and Self: thus, *Maheshwara* means the
great lord and great ruler of all.

Ishwara consciousness arises as an expression of *Maheshwara*
in creation. The awareness that is the formless *Mahesh* is also in
form as the Lord of Creation *Ishwara*. Just as a ray of sunlight is
not different from the sun, and a drop of water is not different
from the ocean, so is Maha-Ishwara, God and Self in infinite
flow: the creator and you in eternal union.

As Krishna shares in the *Bhagavad Gita*: *"Ishwara dwells, O*
Arjuna, in the heart space of all beings." Ishwara is the Lord in the
heart, illuminating the heart of every being, thus Maheshwara is
"the Great Lord who lives in the heart of all beings", which is your
true Self. This is known today as quantum resonance, which
means the same vibration in me is found within you and in all
beings, from which the greeting "Namaste" comes.

In Saivite teachings there are five faces of Shiva: Creation,
Sustaining, Destruction, Veiling and Liberation. The name of
Shiva as Maheshwara is his *fourth face*, a mask that veils and
obscures reality from us in *tirodhana Shakti*. This fourth face of

Shiva hides, veils and cloaks the true nature of reality from us as a necessary part of the whole creation process.

The Maheshwara Sutra is the clear Revealing of Shiva's fourth face. This fourth face veils the formless awareness from where all the processes of creation, sustaining and destruction arise.[18] In this masking or *tirodhana Shakti*, one becomes immersed in and identified with the world, with our experiences and memories, and the "I" that is generated from this. We become this "I", forgetting who we are beyond this limited identity in the fourth face of Shiva Maheshwara.

All these processes are systematically named throughout the Sutra in the 14 classes of sound and 42 syllables that create the human being and universe. The Sutra shows this matrix of all creation, from which all possible processes and experiences arise. In non-dual Saivism, these experiences and actions are like layers of clothing that we put on, and take off. When we let go of identifying with these layers, then That which never leaves us is revealed in Shiva's fifth face: liberation and pure awareness.[19]

The word *"Sutra"* is similarly multidimensional to the word *Maheshwara*. A *Sutra* is a thread and string, the warp and woof of a fabric,[20] a gushing font of teachings, practices, meanings, or a stream of chanted words.

A Sutra is a technology of consciousness, through which an entire body of knowledge is encapsulated in seed forms that are threaded together like jewels on a string, each one leading to and explaining the other. A Sutra is a thread composed of many seeds that string together pearls of wisdom.

These threads can be seen in modern terms as superstrings. As Professor Michio Kaku shares, "the subatomic particles we see in nature are musical notes on a tiny vibrating string. Physics is nothing but the laws of harmony that you can write on vibrating strings. The universe is a symphony of vibrating strings."

A Sutra is a holographic system of data and wisdom storage and retrieval accessed through expanded states of consciousness. Sutras are codes that evoke pictures, meanings, light frequencies and equations, having up to seven layers of information within them.

A Sutra unfurls and is comprehended through contemplation and meditation, revelation and devotion, concentration and insight, uniting scientific and meditational processes with an artist's eye. Sutras are decoded and understood through utilizing both sides of the brain: the rational and the intuitive, the structured and the spontaneous. To understand and create Sutras one has to unite the linear and spatial, the alphabetical and pictorial, the masculine and the feminine, the logical and poetic: art, soul and science.

Sutras and equations both express living, moving relationships. Poetry in sutra is the most concise and multihued form of language, just as equations are the most succinct form of expressing the mathematical processes they describe. Many mathematicians often describe formulae in terms of beauty, elegance and emotion.

As Professor of Neurobiology Zeki at University College London says: "To a mathematician an equation can embody the quintessence of beauty. The beauty of a formula may result from simplicity, symmetry, elegance or the expression of an immutable truth. For Plato, the abstract quality of mathematics expressed the ultimate pinnacle of beauty."[21]

Zeki discovered that the experience of mathematical beauty correlates with activity in the same part of the emotional brain as the experience of beauty from art or music. Equations, music, art and poetry all interconnect through the beauty we feel in them. Beauty is in the eye of the beholder, whether it is abstract, intellectual, soulful, sensory or sensual, or whether it is through a sutra or equation.

Equations, like emotions, transcend the boundaries of language, and both are universally understood languages with a universally understood beauty to them, just as the language of sound and music has a universally understood beauty to it.

Indian-American astrophysicist Chandrasekhar remarked that when he discovered any new insight, it was something *"that had always been there and that I had chanced to pick it up"*. The laws and equations of the universe have always been here, independent of human existence, so in one sense scientists are like cosmic archaeologists, trying to discover and reveal what is already here.

This is similar to the laws of mantra and sutra, which are vibrational superstrings of sound currents and universal processes. Certain mantras and sutras already exist, weaving vibratory currents throughout creation, and many of these have been seen/heard and shared to others (like the Vedas) by sages and seers. All we need do is tune into these currents of mantra and sutra, which are like "super-string" links between dimensions.

The Maheshwara Sutra has mathematical precision like an equation, and all sutras have laws for their efficacy to transmit their wisdom in the most concise, efficient and elegant manner.

The compacting of multiple meanings into seeds in a sutra (known as a holoplex in contemporary terms) can be decoded, revealed and explained by one who has the capacity to move in all four modes of sound consciousness.[22] Such a person unfolds and blooms the superstrings of vibration that compose a Sutra into full understanding and expression.

The laws for the composition of Sutras are:

1. *Svalpaksaram* – It must have the minimum number of syllables. Not even one syllable should be extra or superfluous.

2. *Asandigdham* – There should be no doubt or ambiguities. There should be clarity in essence.

3. *Saravat* – The Sutra should have something worthwhile, of value and importance, to express. It should be powerful, expressing the essence of the subject, and be full of meaning.

4. *Visvatomukham* – It should have a wide applicability in diverse situations, and not be confined to a few particular instances. It should be universally applicable.

5. *Astobham* – It should be free from inadequacies, errors and fillers. It should stand on its own strength and flow freely.

6. *Anavadyam* – It should present an irrefutable truth and be easy to pronounce, even poetic. It should be "agreeable" in its understanding and conveyance.

Does this not sound like laws for an equation? As equations are a foundation for physics, so sutras are a foundation for scientists of vibration and sound consciousness.

The Fourteen Classes of Sound

अत्र सर्वत्र सूत्रेषु हयन्त्यं वरणं चतुर्दशम्।
धात्वर्थं समुपादिष्टं पाणिनियादीष्टसिद्धये॥ २॥

Atra sarvatra sutresu
antyavarna caturdasam
Dhatvartham samupadistam
Paniniyadistasiddhaye

2

Here and in all the verses throughout the Sutra,
Panini's perfected ability perceived
that the last letter of each of the 14 classes of sound
is meant for knowing the hidden meaning of the root word.

This last letter defines the root word,
explains the root word,
thus, bringing together all the meanings of the root word,
which then blooms all meanings into manifestation.

This enabled Panini and other learned Sages to learn the
science of audible and inaudible sound,
as desired by them.

Sanskrit grammar has been called the "royal road to liberation", "that which explains fully every activity", "the supreme essence of speech which assumes form", and the bridge "to pass beyond the duality of light and darkness".[23]

The master of Sanskrit grammar is Panini, who through his perfected power of perception or *siddhi* elucidated the structures that underlie the creative processes of Sanskrit. His comprehensive phonetics, phonology and morphology show Sanskrit as a refined, sculpted, elegant, complete and "perfect" language, a language which the gods clothe themselves in, and through which they are revealed.

Panini was present to receive the Maheshwara Sutra and its wisdom, bringing the sounds into the form, structure and shape of Sanskrit in his *Astadhyayi* or Eight Meditations, in which he revealed the linguistic architecture, code and grammar of the Sanskrit language, whose general "shape" hardly changed for the next two thousand years. He also composed a treatise on *Siksha* – marrying the science of sound and grammar together.[24]

The last letter in each of the 14 classes of sound is a consonant. Consonants create forms, which is what the Sutra details: the forming of creation from the formless. Each class of sound is defined into a set shape, boundary, structure, container and vibratory waveband by the consonants at the end of each class of sound. Each consonant guides the previous instructions of the vowel sound code in order to coagulate energy into forms.

Each consonant is preceded by a vowel. Vowels are *svara*, self-luminous, made of light, and each vowel ripples forth waves. In quantum physics, waves of light do not travel. They mirror each other from wave-field to wave-field of space. The planes of zero curvature which bound all wave-fields reflect light from one field to another.[25] Each reflects the other.

Vowel waves need consonants to form edges, thereby unfolding forms through the refraction of light, creating multitudes of angles and meanings. The consonants provide the code or boundary for the wave of vowel light, bouncing and reflecting the vowel through its harder sound. Each consonant is an angle reflecting the light wave of the vowel sounds.

Each of the 14 classes of sound has its own unique laws, geometries and operating system that is its function, all of which defines its form. Each of the 14 classes of sound are a wave-field of consciousness made up of its unique sounds, creating a pulse wave and vibratory power.

Each of the 14 classes of sound are a set category of vibration within which a specific function of creation occurs, and each class of sound connects to all the other classes of sound in a reverberating vibrational matrix.

Each of the 14 Classes of Sound are 14 different states of consciousness or 14 different dimensions. Dimensions are not places or locations; they exist everywhere at all times. All dimensions are here right now, right where you are sitting!

All dimensions are states of consciousness vibrating at specific rates, similar to a radio, where you can tune into any "station" or vibratory rate of your choice. If you look at a keyboard, imagine each key to be a different dimension. We access each dimension and class of sound by using its specific sounds to enter it, shifting our state of consciousness into that wave-form and category of consciousness.

Each sound has a number, as mathematics is an outcome of linguistics in India, and Panini's programming code of Sanskrit Grammar is based on algebraic mathematics, the same maths that fuels Artificial Intelligence. This ancient and sophisticated code can be used to specify computer languages.[26] However, the true "intelligence" part of AI is in semantics or meanings, which is one reason why AI applications such as sentient cyborgs are not yet able to think, discern and connect events in space-time like we do. This will soon change with developments in technology and quantum computing, and Sanskrit grammar could be a key to this.

The programming code of Sanskrit consists of root words or *Dhatus*, the subtlest elements, building blocks, seed essences and irreducible core vibration of all Sanskrit words. There are 2012 dhatus, which in the process of *sandhi*, or joining together, create all Sanskrit words.

For each of the 2012 dhatus there is a *Dhatvartha*, an action that emanates from the seed word *Dhatu*, blooming it out into its full series of actions and meanings, expanding it into more complex forms, thereby fulfilling its purpose. For example,[27] the dhatu *stha* has *dhatvartha gatinivrrtau*, meaning that words arising from this root are about stability, solidity and the end of movement. So, in English we have the words stand, stable, station and state.

All sounds have rules to their construction, a grammatical code underlying their ability to create forms. These linguistics principles underlie most agreements and understandings between us, as most classes of mental activity depend upon words. Grammar is therefore known as the purifier of other classes of activity, refining, directing, structuring and particularizing in order to bring understanding and clarity to all expression.

Grammar is a revealer of knowledge and a constructor of inner and outer worlds, echoed in Einstein's assertion that "the mental development of the individual and his way of forming concepts depends to a high degree upon language. This makes us realize to what extent the same language means the same mentality."

Grammar is a programming code. Once programmed, regardless of how the real world operates, the mind will follow these rules, superimposing this version of reality onto the world.[28] In each differing grammar, logic, reasoning, mental clarity and mindset are different, and different forms of humour and expression grow out of the grammar of each language.

Language and its code of grammar can be a tool to help lead the mind into quantum realities, for language structures thought by the meanings we give it. Sanskrit grammar composes the rules of the matrix of reality, as seen by Panini who codified the Sanskrit language from his experience of the Maheshwara Sutra, and lifts one into a luminously clear experience of reality as anyone who sounds Sanskrit has experienced.

A I Un: The Beginning

अकारो ब्रह्मरूपः स्यान्निर्गुणस्सर्ववस्तुषु।
चित्कलाम् इं समाश्रित्य जगद्रूप उण् ईश्वरः ॥ ३॥

*ākārō brahmarūpaḥ syānnirguṇassarvavastuṣu
chitkalāṁ iṁ samāśritya jagadrūpa uṇ īśvaraḥ*

3

The sound *A* is the form of Brahman
without any qualities, attributes or characteristics,
the abiding awareness everywhere.

I is its consciousness,
the animating moving power in all beings.

U is its Self in Creation,
the form and Lord of the universe,
the Assembler and Sustainer of all moving forms.

अकारः सर्ववर्णाग्र्यः प्रकाशः परमेश्वरः।
आद्यमन्त्येन संयोगात् अहमित्येव जायते ॥ ४ ॥

*akāraḥ sarvavarṇāgryaḥ prakāśaḥ paramēśvaraḥ
ādyamantyena saṁyōgāt ahamityēva jāyatē*

4

A is the first sound, the first light, first illuminator of all,
Self-Luminous appearing within the Supreme Self.

With the union of the first and last letters *A and Ha*
is *Aham*, the I Am.

The existing self Aham is generated in all its detail
through all the sounds, from the first to the last.

अकारं सन्निधीकृत्य जगतां कारणत्वतः।
इकारस्सर्ववर्णानां शक्तत्तिवात् कारणं मतम्॥ ७ ॥

*akāraṁ sannidhīkṛtya jagatāṁ kāraṇatvataḥ
ikāras sarvavarṇānāṁ śaktitvāt kāraṇaṁ matam*

7

A is Clear Colourless Light, blooming forth
the ever-moving energies of the universe in *Sakti I*,
the activating, dynamically creative power,
the lustre expressing *A* in all the colours and hues of creation.

Due to the closeness to I to A,
I powers all other sounds
and is the inner potency of all letters,
cause of all alphabets,
the seed for the emergence of all worlds,
the source for all manifestation.

अकारो ज्ञप्तमिात्रं स्यादकिारश्चतिकला मता ।
उकारो वष्णिुरत्यिाहुर्व्यापकत्वान्महेश्वरः॥९॥

*Akaro japtimatram syad Ikarascitkala mata
Ukaro visnurityahur Vyapakatvanmaheshvarah*

9

From *A* arises *I*, the mother of sound,
which measures out *A* in the intelligence of eternal time
through *Visnu U*, the omnipresent awareness,
the all-pervading *Vyapaka Maheshvara*
who spreads His presence throughout Himself,
assuming innumerable forms instantly,
extending everywhere, sustaining everything,
fulfilling the purpose of existence,
and being the very sense of existing itself.

A: Opening

"A" is Being in its simplicity without any idea, modification or quality: the initial Being.[29]

A is the awareness of *"the form and nature of the Creator without qualities."* A has no positive or negative charge, no ideas or concepts, no form or colour, nothing to measure itself against; no subject or object. A is colourless clear light, without qualities: It is the awareness which everything appears in, the underlying awareness which all processes and actions appear within.

Clear Illuminating Awareness A has no form, has no limit, no name and no vibration as we define vibration. It does not move because there is nothing beyond it. It has no boundaries, no "here" to differentiate from "there". It has no beginning and no end, no space and no time.

AH has no form or content for the mind or experience. It is the Illuminator, which means if we sound it, everything in creation can be illuminated and become known to us through this single sound. As Krishna said, *"of all sounds, I am the syllable A."*

A is the first sound and first light, the first vowel sound *svara* leading all others. All Sound has a single source, and this single source is That from which all sounds arise within. From *A*, all sounds are made possible in order to express and experience the inexpressible. A is the Transcendent and Immanent, Singular and the whole.

A is the entire teaching of all sacred traditions in a single sound,[30] the shining light of awareness within everything. A is the gateway to the Infinite, the first Sound of Opening, Inspiration, and Illumination. *AH!* We use this sound to ascend, to uplift into spirit.[31] Yogis and monks from India and Tibet use AH to leave their bodies consciously at the time of their death.

A is the First Movement of Creation: pure joy and Self Delight of the Creator, a joy that has no reason or cause to be joyful, *it just is* joyful. This is the vibration we are created from, the vibration we are always held in, and is our most fundamental vibration.

This movement is the subtlest and most powerful, and many people experience it when making love. When we make love, *Ahh* is the most common sound we make (along with O) as at this moment we open to the timeless present moment. In this Moment we forget our identity, we forget who we are, in a powerful, felt "experience" that is actually the absence of all conditioned experience of who we believe ourselves to be.

AH is the Origin of all the vibratory, wavelike processes of the universe and us. When we live in AH, we live according to our true calling, our inner voice and soul purpose which is part of *Rtam*, or Cosmic Harmony. Living in a way that keeps us connected to our core of *A* leads to joy, which is always here – all we have to do to access it is to let go of anything standing in the way.

A is our origin, and how we have been created: in joy and delight. When we feel this presence, we express divinity. As with any true mantra, we do not create it, for the sound is already here. We can tune into this vibration that is always present by sounding it in meditation in the crown, heart and the soul star chakras above the crown chakra.

Sounding A

akārō vai sarvā vāk saiṣā sparśōṣmbhirvyajyamānā bahvī nānārūpā bhavati

A is the whole of speech, and manifesting through different kinds of contact (consonants) and of wind (sibilants), it becomes manifold and different.

– Aitareya Aranyaka 2:3:6:14

A is the awareness of *"the form and nature of the Creator without qualities"*, which unfolds through *I Un*. *I Un* then unfolds into the subsequent thirteen Classes of Sound of the Sutra, showing the unfolding and evolution of the universe from the un-manifest to the material world.

A I Un are the root vowel sounds common to all languages, the most identifiable of all sounds to all humans, the most natural sounds we, and babies, make. From this trinity, all other sounds are derived.

These three vowel roots are known as *svaras* in Sanskrit, meaning that they exist and shine by themselves as pure light. Patanjali shares that a svara is self-luminous and does not depend on anything else for its existence: it stands alone.

Vowel sounds require the least effort for our mouths to make, with no tongue or teeth involved. Each sound flows into the other in one continuous fluid movement, and together are the simplest, most powerful sounds of creation.

This trinity of the primordial vowel sounds *A I U* have been used worldwide in many sacred traditions in order to open the gates of perception in meditation, and in ceremony at the potent times of twilight and dawn, the "gap between worlds", when access into the doorways of perception is easiest.

When sounding the vowel sounds, be aware of the sound with tender gentleness. Become aware of where it comes from within you, and where it goes to when it fades away. Listen deeply to the silence after the sound. Listen to the subtle tones of breath, space, sound and silence behind the sound.

Rest in the space in and after the sound. Know there is space within the sound too. *"All syllables are born in space, resonate in space, and then melt into spaciousness. Know this spaciousness as yourself."* The shining light of space, its glow, permeates your body in all directions. Space is already here as this presence.

Before you sound AH: relax. "Tune" into this ever-present sound, *and allow AH to sing through you*. To sound AH, open the mouth and throat and gently exhale, allowing the sound to come forth effortlessly. Allow the sound to be long, soft and gentle.

Let the sound and breath carry you on its sibilant stream. Gradually, the sound will lead you into light flow and silence, as all sounds and their meanings, forms and functions can only be truly known when we are in silence.

In our human experience, sound leads into light when we focus on the sound. The sound becomes a subtle wave current of light in the brain and body, and then becomes a field, which leads one into silence, giving one the direct experience of the sound current one is meditating with.

Without thought, without image, open, relax into, and rest in this expanse.

Focus and meditate:
Above me is endless space: AH
Below me is endless space: AH
Behind me is endless space: AH
To my left is endless space: AH
To my right is endless space: AH
Within me is endless space: AH

A + I

The first sound AH arises from 0, where there is nothing and everything. 0 is in every other number and in everything, and is the awareness from which all vibrations, all sounds, all consciousness and all experience appears.[32]

In this void there is no motion, so SOMETHING has to move, has to be created; so, 1 arises from 0 as a perfect sphere. This first sphere of creation *A* replicates itself through the energizing sound *I*, forming the Vesica Piscis in this dance between 0 and 1.

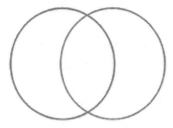

The Vesica Piscis contains the three most important square root harmonics of 2, 3, 5, and the potential for the geometric laws of light. The Vesica contains the primordial shapes of the triangle, square, pentagon and hexagon, and[33] from here Phi or the Golden Mean, the infinite number of 1.618, births, to then weave its spiral nature throughout creation, maintaining the harmonic unity of creation in all forms.

The unfolding of the first two spheres of creation *A I Un* is described in the *Rk Veda*: "In the beginning, pregnant darkness was by enfolding darkness secretly enfolded; unformed, unseparate, that fluid was this entire creation. While the boundless source of Being was by unformed Being thus Enclosed, That One, through light of Knowledge, brought itself to Be."[34]

A and *I* then form *Un*, the "space" or movement between the two spheres. This is called reversal-transference, where the two complementary forces of *AH* or 0 and *I* or 1, in the presence of the third stabilizing factor of *Un*, exchange modes and One becomes the Other. This is primary magnetism and compression. The centre of *AH* or 0 expands exponentially through *I* multiplying this energy.

In *A I Un* the quantum field appears via the Vesica Piscis as its womb. Every subsequent movement in creation is based on this indivisible unity of *A I Un*, the first trinity of creation and the sounds of the quantum field.

A I Un is the Trinity of Shiva-Shakti-Visnu/Isvara. You are the third part of this trinity as the Individual, a particle of the formless in form, the Lord of Creation in creation. Each of us is this Isvara, a living part of God in creation, as Jesus shared, "Ye are all like gods." Shiva-Shakti is in every person and every part of creation. You are not separate to the "Holy" Trinity: you are part of it.

"A" is Brahman, without qualities, Clear Light. Only Awareness is aware and only the infinite is aware of the infinite.

"I" is Shakti, the energizing, activating consciousness of *A*, the Breath of Brahman. Only Consciousness is conscious.

"U" is Visnu/Īśvara, the sustaining of the sense of existence.

I: Activating

Shiva or Awareness *A* replicates itself in the second sphere of creation to appear to Itself as its Consciousness *I*. *I* is the consciousness of *A*, its animating, moving power, the "Breath of God" *Sakti*, the dynamic creative energy that expresses *A*, bringing forth the colours and hues of creation.

I is the Sound of Shakti, the desire to create, the first movement extending itself from Pure Being, creating the womb of the Vesica Piscis. Sakti is the universe itself and is not separate from Shiva.

Siva and Sakti are not aware that they are separate. They are interconnected just as fire is one with heat. Sakti is the whole universe, which issues forth from Shiva.

– Abhinavagupta

I moves as the wave of life, animating and generating all creative activity or *iccha kriya*. It is the energy-desire in all action, the power source, the engine for ALL manifestation, the seed-sound for the emergence of all worlds, birthing the possibility of form out of the formless.

I is an intense state, perpetually birthing everything in every now moment all the time, making all things anew in every moment. *I* powers all sounds and is the potency within all sounds, the animator and creative cause of all language, vibration and frequency, the mother of sacred sound, measuring out *A* into eternal time and the sense of existing itself in *Un*.

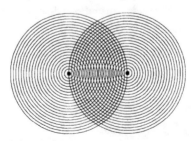

I is pure energy that does not occupy any space, yet drives all activity. I measures out AH as a singular portion or citkala of AH in order to form the second sphere of creation.

जगत्सरष्टुमभूद्वाञ्छा यदा ह्यासीत्तदाभवत्।
कामबीजमितिप्राहुर्मुनयो वेदपारगा:॥ ८ ॥

*"Jagat sraṣtuma bhūdvānchā yadā hyāsīttadābhavat
kāmabījamiti prāhurmunayō vēdapāragaḥ."* 8

"When there arose the desire to create the universe, then the universe came into existence through the letter 'I'. The Seers who Know the Vedas Saw this seed of desire 'I' unfolds the universe through pairs that divide."

In the midst of explaining *A I Un*, Rishi Nandikeshvara shares verse 8 of the Sutra to explain *"I"*, referencing the "Seers who know the Vedas". These Seers knew, saw, heard and sang the Mandalas of the Vedas and this Vedic wisdom affirms and explains the Sutra. Rishi Nandikeshvara is telling us that by looking to the Vedas and their Seers, the Maheshwara Sutra is more fully comprehended.

In *Rig Veda* 9:74:5, the seed of desire *"kāmabīja"* is that which *Soma* (Shiva) lays in the womb (Vesica Piscis) of the Mother of all beings *Aditi Sakti*. This seed sound of Shakti *"I"* unfolds the seed sound *U*, all three of which create the quantum field or universal pattern of creation.

In Vedic Sanskrit, *"bhūdvānchchā"* literally means "one, two, divide". *Kāmabījamiti* is the seed of desire that unfolds the first mated pair of creation *A* and *I*, Shiva and Shakti, which then repeats this quantum pattern, expanding to form the universe.

In the Maheshwara Sutra and in the Vedas, Creation comes from *A*, Awareness, who wished to replicate in order to See and

Experience itself. So, it made a duplicate of ItSelf, a repetition of ItSelf in *I*. In other words, *A* made the One Become the One. This sequence is known as Phi or the Golden Mean Ratio of 1.618, which encourages molecular arrangement through 1 becoming 1, enabling Itself to Experience itself and See Itself from another perspective or angle.

Thus came the merging of the One and the other One into Two. Something progressed with the One and the One Becoming the Two. Yet, the Original One *A* will always remain the Original One or Master Copy, but the other One married the Two and they became the Three in *A-I-Un*.

Then, this 2+3 made 5 in the next Class of Sound *RLRK*, and the 3+5 then made 8 in the fourth Class of Sound *EOn*, and thus was born the first cubic shape of 8, which is 2x2x2.[35]

This quantum code keeps adding further generations of ItSelf, all the way until its ninth division or the ninth beat of Shiva's Drum, as was mentioned in the first verse of the Maheshwara Sutra. (We shall explore this more later on in the book.)

The other sequence or measuring pattern of "one, two, divide", is temporary and mortal. It is of the One becoming Two, becoming four, becoming eight, to then become One again. This is the doubling sequence of how species replicate, grow, mature and die.

This 1-2-4-8 sequence copies itself to fade over time, and has birth and death written into its code. This sexual cell or gamete mitosis pattern is necessary for physical life to live, grow and die, yet is not the quantum, non-dual infinite sequence of *A-I-Un*.

In *Rig Veda* 10:129:04, *"kāmastadaghre samavartatādhi manaso retaḥ prathamaṃ yadāsīt sato bandhumasati niravindan hṛdi pratīṣyākavayo manīṣā."*

"In the beginning, desire arose with the germ of Spirit, and were together the seed-beginning of mind. By the power of mind, the Poet Seers penetrated the heart and found the bond of truth in illusion, perceiving with their heart's wisdom the connection between the existent and the non-existent."

The Poet-Seers or *Kavis* of the Vedas saw that the germ of Spirit *A* together with the seed of desire *I* brought forth creation *as the beginning of mind*. The energy for any creative endeavour is desire, and through harnessing and refining desire, one can pierce the veils of the heart and realize that all existent creation holds *A* awareness or emptiness within it.

In penetrating the heart, one accesses this seed of mind, discovering the truth that lies within the illusions surrounding desire. Our ideas about what love is and what love is not, our appetites, desires and gratification of the separate ego *ahamkar* are what need to be investigated in order to penetrate the heart. Then the Self will naturally shine forth as Delight.

Sounding I

Masculine Names of God have *A* at the end of the word, such as Krishna, Shiva, Brahma, as well as the offerings of *Namaha* and *Swaha* at the end of mantras, all of which raise our consciousness and bring us into the perpetual presence of *A*. Similarly, Kali, Saraswati, Lakshmi and Parvati all have *I* at the end of their names for the Sound of Shakti.

To sound "*I*", as in all the vowel sounds, the mouth is relaxed.

Make *I* without moving the lips, with your tongue level with the top of the lower teeth. Do not vibrate the lips.

Direct the sound through the roof of the palate into the pineal chakra.

"*I*" is potent in the third eye ajna chakra, the cave of the brain, the *muladhara* root chakra, alta major chakra at the back of the brain, and can penetrate and activate any chakra to its core.

Un: Existence

Un is the third aspect of the indivisible quantum trinity of *A* and *I*. It is the formless awareness of *A in existence*. In *Un*, the universal existence is about to begin. It is just being created.

U is the beginning of universal knowing. In *U*, *A* discovers that the universe is existing in its own nature, full of consciousness and bliss. When this is discovered, another movement begins in the next Class of Sound, where It observes that It wants to find out what is existing in this Self of mine.

Un is the pure intelligence of eternal time from where creation arises from, providing the basis for everything to actually exist. This perpetual "now" moment is not a moment in time; it is an ever-present openness through which all experience and existence flows.

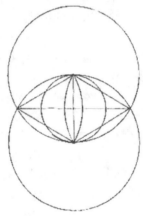

Un is the Sound generated between the two spheres of the Vesica Piscis, the energy between the two spheres of 1 and 1, and now 2. This diamond shaped "eye" forms between the Vesica Piscis of the One, duplicating Itself through *I* and now *Un*.

Un brings forth the next coding sequence of the Phi Ratio and Golden Mean, 1, 1, 2, to now make 3. This sustaining ratio of Infinity duplicates and maintains the Infinite sequencing code of the One awareness. From here, 2 and 3 arise, with the basic shapes of triangle and polygon.

Un is where *A* (formless) and *I* (generator) conceive the possibility of creation. *Un* is the perpetual infinite number and sound current of Phi resonating throughout creation all the time as the omniscience in existence. Everything in creation is upheld by this sound, as *Un* is the quantum blueprint for the processes of creation.

Before *Un*, there is no possibility of anything existing, of any experience being formed, and of any experience existing at all. *Un* is free of all memory that experience brings, and is the Assembler, Sustainer and Operator of *A* and *I*.

Un is the omnipresent awareness of *A* that sustains everything, the all-pervading *Vyapaka Maheshvara* which spreads Its Presence throughout Itself, becoming one with all things It unfolds and therefore dwells in, yet totally free of the form.

Un assumes innumerable forms instantly, extending everywhere, sustaining everything, unfolding the very sense of existing itself. *Un* fulfils the quantum archetype and purpose of creation, how and why *A* and *I* decided to form.

Un sustains the animating desire movement of *I*, creating the actual sense of existing, which is infinite time found in the Phi Ratio Golden Mean. Before this, there is no sense of time at all. With the creation of infinite time, the sense of existence now manifests.

Without *Un* there would be no sense of existence, no continuing of creation beyond Clear Light *A* and its Action of *I*, with no possibility of quantum matter. All of this occurs in the same unbroken continuum of the indivisible quantum field.

In the Sutra, *Un* is the governing *Isvara*, "the unchanging, underlying principle of life."[36] Everything in creation has at its centre an Isvara, a ruling creative principle that measures out the forming and functioning of itself as a created form[37] that is in harmony with all of existent creation *and* the non-existent.

This Isvara governs and sustains the One manifesting throughout creation, and is eternal, free from change, free of judgment and any conditioned way of being and memory. As Patanjali shares, Isvara is "untouched by conflict, ignorance, egotism, desire for pleasure or avoidance of pain, action, reaction and the fear of death."[38]

Isvara is the formless awareness in creation, the Lord of Creation: it states that "this universe is my own expansion." It is your consciousness as an Individual Particle of the One. Isvara is an emanation of Maheshvara or AH, a whole singular part of Maheshvara in creation and manifestation, not separate from Maheshwara. For just as a ray of sunlight is not separate from the Sun, so Maha-Ishvara (the great Ishvara) is the source of Isvara.

In the *Bhagavad Gita*, we are told we can connect to Ishwara through a direct relating with a form of God/dess we feel most drawn to. Isvara is intimate and individual in this sense, which is why Visnu, the Lord of Love and devotion, is named here, as he is also the Sustainer of the universe. (Phi Ratio.)

As the enlightened sage Sri Anandamayi Ma shares, "Where Lord Isvara is manifested, He Himself appears in action, yet ever remaining the Non Doer. Non action *akriya* yet form, *akara!* Every form is He in His own form *sva akara*, the Eternal Self *sva*, revealed as form, *akara*." (Akara also means letter.)

Love blooms forth the creation of existence as well as its end (as we shall see at the end of the Sutra) and creation can never be Known without love. This is one message of the Maheshwara Sutra. *Un* is Living Light, loves quantum resonance in all sentient beings, sustaining the heart in all beings. Each life form has *Un* at its centre and heart.

In Christian terms, *Un* is the Holy Spirit, birthed at the moment the universe was created in order to manifest the blueprint of the One throughout all creation. *Un* is the eternal present moment or Now moment, a perpetually resounding ever-present love, wonder and awe. Love is only truly experienced in the present moment, and is always experienced as new and ever fresh.

Visnu is mentioned in this Sutra as *Un*, and his function here is the same as Shiva in his role of *Vyapaka Maheshwara*, which spreads Its Presence throughout Itself, becoming one with all things It unfolds, extending everywhere, sustaining everything, unfolding the very sense of existing itself.

Visnu is the Sustainer of the Universe, the love that sustains everything. As Rupert Spira shares, *"Love is the natural condition of all experience before thought has divided it into a multiplicity and diversity of objects, selves and others."*

This is the natural indivisible unity of the quantum realm of *AlUn* that *Un* manifests. Indeed, the very word Visnu ends in *U*, a *U* that stretches out endless time and within which all life appears through this omnipresent awareness extending everywhere.

Visnu and Shiva-Maheshwara are the same One expressing different functions. *"Shiva is a form of Vishnu and Vishnu is a form of Shiva. Shiva dwells in the heart of Vishnu and Vishnu in the heart of Shiva."* Bhagavad Gita (Ch. 9, v. 23).

Shiva and Visnu are different names for the differing functions that are played by the One, and each are in the heart of the other. *"Shivasya hridayam vishnur: Vishnoscha hridayam shivah."* The famous Shloka *"Shivaya Vishnu rupaya Vishnave Shiva rupine"* also shares that Shiva and Vishnu are the same form. This is echoed in the form of *Shankaranarayana*, the combined form of Shiva (Shankara) and Vishnu (Narayana). Shankaranarayana is also called *Harihara* – Hari (Vishnu) and Hara (Shiva).

In the direct experience of Visnu and/or Shiva, one realizes they are indivisible. It is undeniable and can only be experienced in your own heart and consciousness as true. In the Sutra, the Visnu syllable *Un* or *unmesa* is in indivisible unity with the *A* of Shiva in the quantum field.

The functions of *A* and *U* are the same yet subtly different in order to help form existence, so the One of *A* and *U* can enjoy the experiment and experience of love in form. *A* and *U* are the basis of AUm – the sound in all creation.

Even the two different syllables *A* and *U* flow effortlessly into each other when you say them: *AH* starts with open mouth, and as you close the mouth by pursing the lips, *U* is produced. No other movement of the mouth, tongue or teeth is required. *A* and *U* are the two most common sounds in all human languages, and the vowel sounds that are the basis of all languages.

Sounding Visnu-Shiva: Hari-Hara

In spiritual practice, the combined name *Harihara* is a powerful breathing practice uniting the Hara and Heart together in One flow.

Inhale HA into the Hara
Exhale RA out from the Hara
Inhale HA into the heart
Exhale RI from the heart
And continue x108.

Sounding Un

A I Un is the quantum field trinity of Clear Light and Energy forming everywhere. It cannot be divided as it is One Presence flowing as a wave, pulse and field of joy, aliveness and spontaneous bliss that has no reason to be blissful – it just IS blissful. This is the basis on which the universe is formed;

Infinite Being overflowing through pure joy and potent desire energy to form eternal time and existence.

If you sonically imagine the sound *U* or *Un*, it is an infinite sustained tone, going on forever. Tune into this vibration that is already here, and merge your sounding into it to enter this Sound. It is waiting for you to be still and feel-hear it, for this Presence is resounding around and within you all the time.

U is a wide, extended but not diffused state of existence. U is the first of the labial (lip focused) family of sounds, and leads them all. Start with your mouth open, heart relaxed, sounding "*A*", and then narrow your lips as if blowing out a candle: "*A*" will turn into "*U*". It is part of the same movement. Purse your lips until they are closed and continue the *U* into *Un* to vibrate your pineal.

Un is a sagittal sound, uniting the central planes of the brain in the corpus callosum and pineal. It especially resonates with the alta major chakra at the back of the head, and continues from here up and out of the head in a diagonal axial line through the pineal. It also resonates well with the chin chakra, the throat chakra and the heart chakra. Try sounding UN for a few minutes at a time in each of these chakras, starting off with the crown.

CROWN
PINEAL
ALTA MAJOR and diagonal axial line out of the head
CHIN
THROAT
HEART

Sounding A I Un

The first sounds of creation *A I Un* are the clear light of formless awareness, desire, and their union, which is eternal time existing in love. This is the basis of the universe, the basis of creation, and exists everywhere in all sentient beings. What a wondrous creation!

A I Un can be sounded in our brain chakras, as the Vesica Piscis marks out the left and right sides of our brain, with the corpus callosum and pineal chakra in the middle. You can use the vowels *A I Un* as they are, or add the *Anusvara MMM* onto the end of each vowel.

So *A* becomes *AM*, which are Tantric sounds that can be used in the soul star chakra, crown chakra and palate chakra. *I* becomes *IM*, which is the original sound of AUM in the *Rk Veda*, and *UN* already has *n* in it.

In Tantric mantra we add the nasal humming sound *MMM* to seed syllables in order to bring the formless into form. This activates the pineal and pituitary glands in the brain. NN and MM modulate the vibration of the vowel to bring it into form (as we saw in verse 3 previously). Try both ways, with just the vowel and then adding the form-making consonant, and see what resonates with you the most.

Anusvara or *"MMM"* is a dot above certain Sanskrit letters such as *Aum*, and is a gateway into the Soundless Sound, *Paranada*. The *anusvara* vibration gradually dissolves sounds into silence.

Anusvara is a bridge, leading from non-being to being; from sound to silence and from silence into form. Anusvara is a point of pure Illumination (*prakāśa-bindu*) an experiential awareness that re-enters into the Absolute [for a moment] before projecting out into manifestation.[39]

To sound Anusvara make the *"MM"* sound *without closing the mouth*. If you close the mouth MMM will resonate the lips instead of the pineal chakra. The MM vibration should come from the upper palate and the nasal passages.

When you do this, your cranium vibrates instead of your lips. After you can make the Anusvara sound *MMM* without vibrating the lips, then you can close your lips when you say it, but make sure that the lips are NOT vibrating. In certain Tantric schools the *MM* becomes *NG*. Try both and see which one resonates most for you.

A I Un can be used in all chakras as the quantum field exists in everything. Here is one example out of many you can use *A I Un* for:

1. *AH* – Opening Inflowing Light – Soul Star Chakra, Crown, Palate
2. *I* – Shakti Activate – Ajna Chakra, Medulla
3. *UN* – Soul Star Chakra, Throat

A I Un Body Prayer

Focus on your soul star chakra, one hand's width above your crown. Hear *AH* as an infinite sustained sound, then join in with this *AH*, out loud, and then internally sounding with it. Have your arms stretched out wide at the level of your heart, with a wide-open heart.

Bring your hands together, all fingers interlocked with your first index fingers pointing up, at the level of your pineal. Hear *I* as an infinite sustained sound, then join in with this *I* out aloud, then internally sound *I* in the pineal.

Bring your palms together at the heart. Hear *U* as an infinite sustained sound, and then join in with this *U* out loud, then internally sound *Un* in the throat and heart.

These are just guidelines. Allow the sounds to guide you. They are infinitely intelligent, and will take you where you need to go, and inform you through your feeling intuition where to sound them.

A I Un in Geometry and Cymatics

Cymatics is a modern science that records what sound looks like as geometry. Every sound and vibration (*nama* or name in Sanskrit) has a geometry, a form to it (*rupa* in Sanskrit), and both sound and form *are the same*, indivisibly intertwined in the creation process known as *nama-rupa*. *Cymatics*, meaning wave in Greek, captures what sounds look like.

As you can see in these three Cymatic images of *A I U*, they are closely related, one unfolding from the other in a sequence.

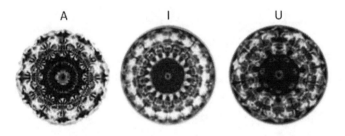

A is a full and complete geometry with 12 petals to its yantra, 12 being a masculine number of order and structure. *A* is followed by *I*, its consciousness, which shows a going inwards, a refining, defining motion after the expanse of *A*.

I has 18 petals, 18 being a harmonic of the *feminine* number 72 and an angle of the feminine shape of the pentagram.

U expands and blooms outwards from *I*, like a flowering petal coming forth in more fluid and complex structures, creating a 7-fold geometry, 7 being the number that denotes development or evolution *in time*.

AHAM: I Am

The existent Self AHAM is generated in all its detail through all the sounds, from the first to the last.

In the *Puranic* period the Sanskrit Alphabet was called the *Ahaà*, representing A to HA. *A* is the first sound of pure light, the Illuminator of creation, and *HA* is the last sound of the Maheshwara Sutra alphabet.

HA is the completion and dissolution of all sounds, when all vibrations and forms have emerged, when all creation has occurred, and the Intelligence behind it all dissolves from its creation in the supreme detachment of pure love.

This is a simply put revelation: the 42 sounds of the Maheshwara Sutra are what the universe is made of and the full range of experiences available to the Self.

The sequence of A to HA contains every phase of consciousness and all the elements of experience in our inner and outer worlds. This wave, field and particle energy reveals the quantum field and space-time, and connects all the different realms of creation.

A and HA are the first and last sounds of the creative process, the beginning and the end, the Alpha and Omega. AHA is the sound of inspiration, when we have Realized or understood something revelatory. When we add "MM", the Anusvara sound, AHAM arises.

The entire Maheshwara Sutra shows this single awareness AHAM I AM as being ever present and behind, yet not dependent on or attached to, the movements, experiences and meanings of creation.

The name we give to that which is aware or knows about experience/creation is "I". I is the Knowing element in all experience, the aware element within all experience and creation. I refers to that element of our experience and creation that remains consistently present throughout all experience and the unfolding movements of creation. Therefore, I is the common factor in all experience/creation. What element of experience is always with you, has never left you, has never changed, throughout all your life? The I AM.

Creation is continuously evolving, changing, appearing and dissolving as the Maheshwara Sutra shows. I AM remains consistently the same and Present throughout all these changes, appearances, forms and creations. I AM Knows all experiences, movements of creating and all forms created, but is not itself an object or movement of experience.

I AM is the knowing element that accompanies all experience, but is itself not an experience. All objects of experience are known by this I AM, but I AM is only known by Itself. No object or form can know it, no experience can know it, only It Knows Itself.

This being that shines in each of our minds as I AM is not a personal being or self. It is the single, infinite, indivisible impersonal being, refracted into numerous apparent selves without ever becoming fragmented. We all share the same being.[40]

Sounding AHAM

There are many ways to resonate, connect with and enter AHAM. *Here are 5 different sonic ways.* Try them all, and see which one takes you deepest into AHAM.

When you enter the sound field consciousness of AHAM, let it take you. The sounds you make may dissolve into light or deep meditation, so allow this to happen. Have a gentle and relaxed focus when sounding AHAM.

Remember that this sound AHAM is already ever present here and now in creation, and indeed is only available and accessible in the Now Moment. AHAM is within you. All you are doing is accessing what is already always here.

When you have entered the field and sound, let it take you. The initial discipline of breath/counting is just to entrain you into this sound field. When you are entrained, let go into it!

1. Inhale *AH* into the ajna pineal for 7 seconds, making *AH* within the breath.
   Exhale *AHmm* whilst focused on the ajna pineal chakra for 7.
   Repeat this cycle at least 7 times.
2. Inhale *AH* into the sahasrara crown chakra for 7 seconds, making *AH* within the breath.
   Exhale *HAmm* focused on the anahata heart chakra for 7.
   Repeat this cycle at least 7 times.
3. Inhale *AH* into the anahata heart chakra for 7 seconds, making *AH* within the breath.
   Exhale *HAmm* focused on the sahasrara crown chakra for 7.
   Repeat this cycle at least 7 times.
4. Inhale *AH* into the anahata heart chakra for 7 seconds, making *AH* within the breath.
   Exhale *HA* whilst focused on the sahasrara crown chakra for 7.
   Make the sound *"MMM"* for 7 focused on the ajna pineal chakra.
   Repeat this cycle at least 7 times.
5. Inhale *AH* into the anahata heart chakra for 7 seconds, making *AH* within the breath.

Exhale *HA* whilst focused on the vishuddha throat chakra for 7.

Make the sound *"MMM"* for 7 focused on the ajna pineal chakra.

Repeat this cycle at least 7 times.

After the 7 cycles, you can go deeper by repeating the sound-breathing internally. You can also extend the counting and inhale and exhale the sounds for 9, 12, 16 or longer.

The Four Modes of Sound

सर्वं परात्मकं पूर्वं ज्ञप्तिमात्रमिदं जगत्।
ज्ञप्तेर्बभूव पश्यन्ती मध्यमा वाक् ततः स्मृता ॥ ५ ॥

Sarvaṁ parātmakaṁ pūrvaṁ jñaptimātramidaṁ jagat
jñaptērbabhūva paśyantī madhyamā vāk tataḥ smṛtā
5

In the beginning, the whole universe was only in *Parā*,
without beginning or end, unmanifest.

From Para
arise the processes of the ever-moving vibrating universe,
measured out in the Seeing-Perceiving of creation Pasyanti,
unfolding through the middle state of sound Madhyama,
into Vak, speech that can express the Supreme.

This is the lore of the lineage.

For thousands of years the study of sound and the ever-moving processes of vibration that create, sustain and unfold the universe in all its aspects and expressions has been a major part of all Vedic schools of learning.

Sound and language was a cross-disciplinary affair, with each school of learning recognising the importance of sound vibration and incorporating their understanding of it into their own philosophies and experiential practices.

From yoga to grammar, from philosophy to sound yoga, from health and healing to meditation and Self Realization, each school of learning had its own understanding of the four modes of sound, leading to a rich and varied tapestry of wisdom about how sound and vibration work across the entire gamut of human experience.

The four modes of sound are first mentioned in the *Rk Veda* Hymn of Creation, the foundation of Vedic creation wisdom: *"Speech is measured out in four worlds. Three hidden in the deepest cavern cause no motion. Mankind gives utterance to the fourth."*

We use the fourth mode through normal speech and all sounds we hear, as well as words and writing. This is what we normally think of as sound, but in reality *is just one of the four modes of sound vibration.* The other three modes of sound are "hidden" from our everyday mindset and waking state of consciousness, beyond normal audible sound and words.

In the four modes of sound described in Nada Yoga and *shiksha* (science of sound), sound is both the source of matter, and the key to become free from it. As the *Bhagavad Gita* says: *"One who thoroughly understands the four modes of sound can utilize this science to become free from the bondage of matter."*

This becoming "free from the bondage of matter" means we are not limited to the body and the laws of our space-time universe. When we understand and directly experience within ourself that matter is made of vibrations, and we experience the four different modes of vibration that underpin this, we have the ability to see and therefore potentially change the vibratory rate of any object, including ourselves.

This means that we can vibrate in and out of different dimensions of existence, as many stories about the Rishis and Masters show us. This freedom to escape our bodily limits is commonly experienced in meditative practices, such as, but not limited to, astral travel and remote viewing, and in new technologies like virtual reality.

As the *Bhagavad Gita* continues, *"Sound conveys the idea of an object, indicates the presence of a speaker, and constitutes the subtle form of space."* Quantum physics now confirms that it is from vibrating waves that all objects appear; waves vibrate the appearance of objects into existence. Objects are actually waves, and it is not matter vibrating but the wave itself that is vibrating.

Without sound vibration, there are no objects. Sound vibration gives rise to the form of every object. Every sound has a form and every form has a sound. As the Vedic texts tell us, when one is mentally clear and in a state of silence, and then looks at any form, the sound of that form will become clear to you. This is called *ritambhara pragna*, which means you perceive forms as sounds.[41]

Sound conveys the idea of an object by bringing many sounds together in groups, known as a harmonic cluster of frequencies. These groups form through waves meeting each other in their different wave planes, and reflecting off each other. These points of intersection are where the harmonic clusters congregate, and then our senses perceive objects from these clusters.

Everything in creation passes through the four stages of sound vibration, and some of these vibrational patterns then manifest into the third dimension to become a physical object. As the *Bhagavad Gita* continues, "sound... indicates the presence of a speaker," which arises through the element of space. Space is the foundational and fluid element in which sound first arises, which forms our capacity to hear and speak. Space is the medium for all waves to move through, and waves are the medium for sound vibration.

All good composers know that the "space" between notes and musical phrases creates a certain atmosphere and feeling, and is as important as the notes of the music itself. This is mirrored in mindfulness practices, where one rests in the space between thoughts, and the space between breaths, in order to enter a meditative and peaceful state.

Similarly, mantras used by hundreds of millions of people worldwide everyday focus on the space and silence after the mantra, where you rest in this space and silence in order to enter deeper meditation and *access the next higher mode of vibration.*

The Four Modes of Sound

Everything in creation is a musical measure arising from *Para,* the Supreme, and in the Maheshwara Sutra, through Shiva's *Damaru* Drum. These musical measures come into manifestation, forming the vibratory qualities and movements of life, through the four modes of sound: *Para, Pasyanti, Madhyama* and *Vaikhari.*

The transformation of the first, Supreme and unmanifest *Para* into the most concrete and fourth mode of vibration in the *vaikhari* of speech depicts the evolution of the Universe from the un-manifest quantum world into our material 3D world.

All modes of vibration arise from *Para,* the Supreme or "the beyond". This first soundless, silent being gives rise to and generates all other vibrational states, sounds, and frequency movements. Para and *Para-vak: the voice beyond* or *Supreme Voice* is the first "Word" of the Creator, the first vibration from which all forms, sounds and functions unfold.

Para Vak or Supreme Speech is a telepathic transmission and imprint of The Word from Supreme Consciousness, a beginning-less and endless state of being that is the basis of creation. Para births the three other modes of vibration through *Para-spanda* (the supreme pulse wave), the constant humming vibration within all life, a vibratory wave flowing through everything in creation, including our bodies and minds, as the Word born from Silence.

This "Word" unfolds itself in its second stage to become the quantum blueprints and infrastructures of creation and perception in *Pasyanti*, literally meaning "Seeing", "Pure Perceiving" or "Visual Sound".

Pasyanti is the first perception or Seeing of creation, where you can perceive anything and everything in creation instantly for what it is and what it could be, as well as its true form, its true sound vibration and its true function. When you Know this about any form or object in creation, you have the ability to summon it, change it, and bend it to your will.

In our experience Pasyanti occurs when one instantly Knows something that transforms you in a Lightning Flash of Insight and revelation, known in esoteric Christianity as Gnosis. Pasyanti occurs when you instantly Know something that transforms you, that changes your perceptions, beliefs and life.

Pasyanti is where the Vedas, Kabbalah, Gnostic texts and many Buddhist sutras were transcribed from *Para* in a holographic revelation. In Greece, Pasyanti was known as *hesychia*, where messages from the gods were revealed to humans known as heroes.

Pasyanti is visual sound, a form of instantaneous holographic communication where whole volumes of wisdom are downloaded into a person in a few seconds.[42] Thus, the hymns and mantras of the Vedas were "seen" by the Rishis, and placed into collections of mandalas that are the visual sounds of the Vedas.

Pasyanti is the mode of vibration where word and object are identical, and where division between subject and object does not exist. Pasyanti is a blissfully ecstatic consciousness in its peak state.

When it flows through us, it can produce sounds through our vocal chords that come from Para. When this happens, it transforms ourselves and the people around us.

Pasyanti is a laser like coherent perception of pure consciousness exquisitely and precisely focused, clear and potent. It was accessed by the Rishis by their generating the inner fire of tapas in *tapasya* – deep one pointed meditations over periods of time.

This concentrated, focused, harnessed fire of one pointed will, desire and intention, became a subtle fire refined to such a degree that it can penetrate anything, see anything and know anything. In Pasyanti you come to penetrate and understand something so totally you become it.

One way you can experience *Pasyanti* is by looking at an object or person without interpretation, in total innocence. This is achieved through concentration and relaxation simultaneously. You observe and look at something, but do not see anything. It is pure sensation without any differentiation, without any thought or interpretation of what you are seeing.

You see the whole of something, the essence of something, not its parts. You can look at a person and practise eye gazing, where you look so deeply into their eyes you end up not seeing their form or their eyes, just a white light and then an emptiness or blankness.

As Pasyanti steps down its vibration, *madhyama*, the bridge, interface or translator, unfolds. Madhyama translates vibrational waves into forms accessible and usable to us. It is the bridge between the 95% of reality beyond the senses and the 5% of reality we see, hear, touch, taste and smell.

Madhyama is a bridge between the waking state of consciousness we normally operate on, and the 90% we use in altered states of consciousness. Madhyama is the "mediator[43] that converts vibrations into objects in time and space through the image-making power of the mind." The objective time/space world, the dreamy subjective world of dreams, hallucinations and visions... are examples of the ability of the mind to convert vibrations into wriggles into patterns we call images.[44]

In madhyama, we communicate through the image making power of pictures, shapes, geometry, Reiki chi symbols, vivid poetry and metaphors, all of which convert vibrations into wriggles into patterns we call images. One picture is worth a thousand words, or as Aristotle shares, "the soul thinks in pictures." The capacity to do this bridges different levels of consciousness. Madhyama is where the medicine of true sound healing lies, as do other modalities of vibrational medicine and energy healing.

Madhyama is known in cognitive psychology as the adaptive unconscious, "an interface between the 5% of our brains that act mainly from habit, and the vast pool of unconscious power within us."[45] This interface, according to Gerald Zaltman of Harvard Business School, is mythological and metaphorical in nature, responding to interior images and symbols rather than rational thought.

Madhyama transforms Pasyanti into shape making processes of pictures, geometries and fluid forms. Eventually, this picture feeling takes a subtle form within, and one grasps and understands it. This is *madhyama vak*, a pre-vocal thought-picture in an inaudible form. Madhyama or "middle" is an interface bridge of vibration in the dream state of consciousness.

Madhyama is a state of wordless thoughts – your own inner world. It has no words to it. This wordless pre-language dreamy state is why many people find it hard to articulate their inner world, struggling to put words to this state and their inner apprehension of things. To articulate this state, we have to come into the next mode of sound *vaikhari*, the realm of words and the mouth.

When madhyama becomes conscious and comes out of the speaker's mouth it becomes *Vaikhari Vak*, words and sound waves moving through air. *Vaikhari* is the vocal expression of your inner feelings and thoughts, your outward expression conveying your inner ideas and feelings.

This of course happens all the time every day, and we become particularly aware of it when we struggle to articulate a deep or subtle feeling. These moments occur when our mind lies between madhyama and vaikhari, and by becoming conscious of these moments, we can change our expression, as we will discover later in the book.

Vaikhari is our normal state of speech and audible sound.[46] This becomes a vehicle for the Word or Voice of the Creator when we are aligned with the universal creative process of the modes of sound emanating from *Para*. The fourth mode of *Vaikhari* is the final expression of *Para*. Our speech is designed to act as a conduit for the Word of Para, an echo of the Word in human form of Source expressing itself.

Speech is our most powerful tool to create with, our very first instrument of expression and the origin of all music. Vaikhari completes the universal hologram by bringing the soundless sound of Para into manifestation through us. Our speech and sounds we make are the end result of the universal vibratory creative process.

The Vedic Rishis understood speech as quite different to what we understand as speech today. These Masters heard and saw the quantum blueprints of creation in all states of sound whilst situated in the Silence of Para, for it is only from Silence that all modes of sound can be fully Known.

THE 4 MODES IN THE BODY

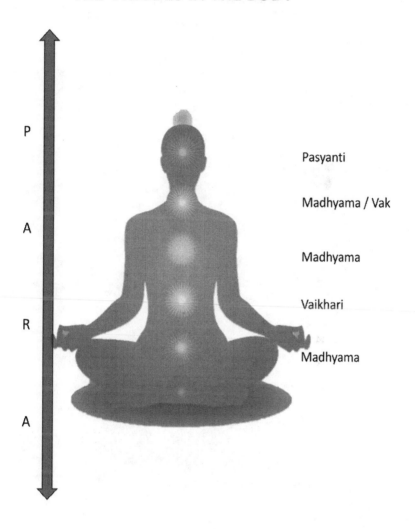

P

A

R

A

Pasyanti

Madhyama / Vak

Madhyama

Vaikhari

Madhyama

Transformation of Sound

वक्त्रे वशिुद्धचक्राख्ये वैखरी सा मता तत:।
सृष्ट्यावरिभावमात्रात्मा मध्यमावत् समायुतम् ॥६॥

*Vaktrē viśuddhachakrākhyē vaikharī sā matā tataḥ
sṛṣṭyāvirbhāvamātrātmā madhyamā vak samā mata*

6

Upon reaching the throat chakra,
this vibrational process mother's fluid shapes of vibration in
madhyama,
to bloom forth in speech *vaikhari*,
expressing one's inner state, feelings, moods and karmas.

One cannot access this madhyama,
or speak the divine speech vaikhari,
without purifying your words and throat chakra.

By its origin and manifestation
Vaikhari and madhyama mirror each other.

To have an open throat chakra means we can fully express ourselves, allowing our speech *Vaikhari* to flow directly, without filtration or blockages, from *madhyama, pasyanti* and *para*. This is our most natural state.

As Rudolf Steiner shares, "all movements that nature uses in creating are found in the human larynx... becoming audible as sounds, speech and tones. In your vocal chords, sound flows into forms, shapes and geometries, coagulating in rhythmic patterns."

There is a whole science behind *Vaikhari* in Sanskrit, as Sanskrit is a language of sound first and foremost. Just by sounding a Sanskrit word, mantra or verse correctly, one can enter the state of consciousness, action or meaning the verse is talking about. To sound a Sanskrit word correctly requires specific movements and shapings of tongue, lips, nose, teeth and palate, culminating in *vak* speech emanating from the mouth.

Certain sounds also require focusing on different chakras and the nose. Just by making these sounds correctly, spaces within you start to activate and open. For example, in the ancient Indian singing art of *Dhrupad*, the student starts by spending weeks, even months, learning how to sound the syllable *AH*, surrendering to its flow, before they can progress into singing other syllables.

This takes time to master, as vibrational singing is an art and science. When the singer does this, the effect on the audience is enchanting, inducing peace and devotion.

All of this is facilitated by the ability to fully express ourselves through the throat chakra. This is intimately connected to the *granthis* or locks of the chakra system, found in the *muladhara* root chakra, the *manipura* or solar plexus chakra and the *vishuddha* throat chakra. To open the throat chakra fully requires that we work on and open these locks, particularly the throat chakra lock, in order to allow a full flow of energy to course through our system.

Physical yoga *asanas* work on certain levels of this opening, emotional healing and deprogramming lead to greater openings, and the movements of Kundalini lead to the greatest opening. Emotionally, childhood and family trauma aligned with societal repression, shaming and distortion blocks the opening of the throat chakra, as does sexual trauma and repression.

Being unable to speak what you feel out of a sense of unworthiness, not feeling good enough, not feeling capable enough, are some of the signs of a blocked throat chakra. You may have been shushed or shamed as a child. You may have been scared to speak out because of potential repercussions or humiliations.

Perhaps people around you simply did not care what you said, or just laughed it off. Perhaps your parents, and partner, ignored you and you learned to live with it, closing yourself down in order to "fit into" the relationship and to fit into society. Perhaps you have a fear of public speaking. Perhaps you believe that what you have to say is not valid.

There are many reasons for the closure of the throat chakra, and as the granthis show, it is connected to your sovereign empowerment and the opening of your kundalini Shakti. When these granthis are relaxed and free flowing, more energy moves through your nervous system and chakras; this then causes your inner thought-pictures in Madhyama to change, which then reflects in your Vak speech, your use of words, and the feeling tone quality behind your words.

As Steiner continues, "the mass of working parts in our throats and mouths are a precisely constructed mechanism giving your life-force a means of expression to shape the worlds around and within you. Within our throats are held archetypal movements of creation: vast possibilities are inherent in the movements of speech pouring from us, so we can express our moods in a kind of meteorology for the soul."

We can be vehicles for the higher modes of vibration to emanate through our voices. The back of the throat chakra can open up in a way that emanates Pasyanti and even Para frequency. My experiences of this are that when I sing certain mantras, the Sound takes over me; I become a vehicle for the Sound to express from Para through Pasyanti into vaikhari.

One becomes an open conduit for the void to sing through the throat, and this is a powerful, transporting and consciousness elevating experience for singer and audience alike.

All possibilities for creation can be found in the voice, and we can access the potency of Vaikhari through mantra. The universe is formed of mantras, harmonic clusters and sequences of sounds, which through the power of our voices we can tune into, make and emulate, replicating the Sounds of Creation for ourselves.

Certain mantras are mathematically and harmonically perfect, and already exist within you and around you as part of the matrix of creation. By sounding them, you resonate with these sound currents of creation: you are not making them; you are just sounding aloud what is already here.

You are simply sounding yourself into a wave of vibration that is already here and flowing as part of the vibratory matrix of reality. By tuning into the right mantras, you surf these waves that form creation.

Mantras are words of power. They open highways for multidimensional communication, and train you how to navigate and experience the multidimensional universe. They are broadband Wi-Fi for the spirit, "phone" lines that when you "dial" them connect you into whatever aspect of creation, whatever part of you, that you wish to commune with.

Mantras are vibratory remedies giving you the frequencies you need. Mantras open chakras and can disturb the course of disease, as well as prevent disease from happening. Mantras cut through psychological impasses, reveal the cause of deep patterns and show you what you need to strengthen.

Mantras still your mind and body, allowing the voice of your soul to be heard. Mantras reveal their power according to what you are ready for and can handle, as mantras, like sacred languages, work on seven different levels of frequency.

For example: if you are chanting a sound aloud in Vaikhari it produces a certain effect. When you take this sound and repeat it internally in the mind or in a specific chakra, the frequency becomes deeper and subtler. One enters the deeper frequency-meaning or tone behind the words in Madhyama.

Over time the mind quietens its chatter and the sound becomes deeper, until you achieve a one pointed focus on the sound current alone. In this total focus of Pasyanti you penetrate, dissolve into and become one with this sound current. As this happens, the sound may repeat itself without any effort on your part, dissolving the mind and your sense of self into the sound current.

The sound current then transforms into a tangible, palpable current of light in your brain. As this happens, the breath slows down considerably, the heartbeat decelerates, and a signal arises from the pineal and brain glands to secrete hormones. These hormones can be felt as deep, intoxicating currents of ecstatic bliss coursing through you, releasing Soma, the elixir of the gods, the pure bliss power of Samadhi in Para.

As this happens, there is no more "you" left, no mind, no thoughts, no personality, no emotion: just pure light consciousness Pasyanti moving at a very high frequency into Para, beyond light or consciousness, into pure awareness. This is when the chakras can completely open, especially in the pineal and pituitary glands.

This then is how the four modes of sound are designed to operate within us. This is the height of vaikhari when it is connected to all four modes of sound, leading to *transformation through sound*.

Vaikhari and Madhyama

As we have seen, madhyama is the state of sound-thought-picture from where vaikhari speech comes from, and as such they are a mirror of each other. These two work together, and culminate in the third dimension of name/vibration and form through their interplay of the inner thought-pictures of madhyama, and its outer expression in speech vaikhari.

In the substance of madhyama, we come to feel and distinguish that:

Every sound is a word.
Every word has a meaning.
Its meaning is what it does.
What it does depends on the substance it vibrates in.

This wonderful insight from Peter Harrison shows us that each and every sound we make is a word, which has a meaning, from grunts and sighs of pleasure and pain, to eloquent poetry and powerful mantra. The word and meaning are the same. The meaning is what the word actually does kinetically, emotionally, mentally and spiritually within you.

What the sound does depends on the substance it vibrates in. This substance changes depending on which of the modes of sound a person's consciousness is vibrating in, as all three modes (except Para) are subtle forms of matter.

In addition, where the sound is focused from has an influence, be it from the heart, throat or pineal. What sound does can vary greatly, as substances are subtle forms of matter (even space is subtle matter) and every vibration has to resonate within a substance.

As sound already exists all around and within us, pronunciation simply excites it and brings it into manifestation in whichever substance you are in, be it emotional, mental or causal, or in this context the three modes of sound. The substance or field you bring forth the sound in changes the meaning, texture, quality and colour of the vibration; what the sound feels like emotionally may be different to what it means mentally or where it takes you spiritually.

A beautiful illustration of this comes from a story in the Brhadaranayaka Upanishad.[47] Indra, leader of the Devas or Gods, Virocana, leader of the Asuras or demons, and Manu, Lawmaker for humanity, all went to Prajapati Brahma, the First Father of Creation, for teaching. He looked at each of them in turn and said: *da, da, da*: the exact same syllable repeated three times.

Indra contemplated this, arriving at the conclusion that da meant *Damyata* – control yourself – because the gods were self-indulgent and easily stirred to both pleasurable and wrathful emotion. Virocana understood da to mean *Dayadhvam* – be compassionate – as demons were naturally angry and known to torment others and themselves, for which compassion is a natural antidote.

Manu interpreted da to mean *Dana*, or giving, charity and generosity, as humanity was known to be greedy and over-materialistic as a way of compensating for the lack of an inner, spiritual life. Generosity and giving was understood to be one way out of this narcissistic self-centredness.

The same syllable "da" was vibrated in three different substances or states of consciousness – the gods, demons and humans – to produce three totally different meanings and interpretations. These three states of consciousness exist within us all, as humans encapsulate all three realms, in our shadow self/vaikhari, our human self/madhyama, and our divine self – found in pasyanti and para.

When we speak in vaikhari, we can give utterance to the most banal words, the most demonic words, and the most divine words. When we are in the fluid medium of madhyama, the internal pictures we see, emanate and create hold vast amounts of information, for pictures are worth a thousand words. When we are vibrating in pasyanti, word and meaning become one, which can then take us into the silence of Para. Para changes the very substance of our consciousness.

The meaning of each word depends on the substance it vibrates in. How many times have you interpreted another's words in these three different ways, from your vulnerable human self, your fearful subconscious self, or your divine loving self?

Can you recognise when you have spoken in these three ways to others in the last few days? Can you recognise your own thoughts coming from these three ways in the last few days?

Check yourself next time you have an important communication:

Are you coming from the heart, the throat or the pineal?

Are you coming from a demonic, human or divine consciousness?

Are you coming from vaikhari, madhyama, pasyanti, or all three?

Which is the best way to communicate your feelings?

In learning to communicate better, you could politely say to the other:

"I hear you, and perhaps the best way to get your point across is through using different words that come from your heart."

OR you could use a picture or a video you could see together,

OR a form of silent meditation with each other to tune in,

OR a sound, chant, or singing together,

OR eye gazing to feel each other deeply and more vulnerably.

One of the keys to understand Vak and to bring it to its height is found in the Upanishads in the *mahavakya* or great saying:

Sravanam, Mananam, Nididhysanam
Hear, Reflect, Realize

Sravanam or hearing begins with *vaikhari*, the spoken word, which we can take in literally at face value as information, *or* we can truly hear what is being said by another through *madhyama*. This requires deep listening, one of the aspects of madhyama, and the ability to be present with another, not with a focus on replying or getting something from the other for yourself, but just to receive the other through the underlying tones, frequencies and feelings behind their words.

In *mananam* we truly listen-receive the other, and then contemplate what is being shared without imposing our own filters on it. We focus on the tone and feeling of the words, rather than the words themselves. We absorb the sound and be with it.

You become present, without any preconceptions or judgments about the sound vibrations being presented to you. You reflect on what is being said, and then its true meaning arises. Sometimes we may see or understand what is being said to us (when we are in this madhyama consciousness) through seeing pictures of what the other is saying.

Nididhysanam is when we Realize the truth through this process. This can occur when we interact with another, read spiritual texts, or hear truths and feel them beyond the intellect in a felt Realization. Sometimes one insight can change your entire life. This is an aspect of *pasyanti*. When we hear, reflect and Realize on an essential truth, sometimes we become silent: pasyanti enters the silence of *Para*.

Vaikhari

From the soundless sound of Para arises the Perceiving, Seeing and Knowing of creation in Pasyanti. From this Seeing arises fluid shapes of sound in madhyama, the bridge into the third dimension of Vaikhari, where vibrations crystallise into speech and words we give voice and expression to.

śrutirapi "vāgēva viśvā bhuvanāni jajnē" iti
sūkṣhmā vāgēva viśvākārēṇa pariṇamatē vivartatē vētyarthaḥ
– Tattvavimarśinī 05-06:b

Śruti (Vēdās) says, "It is only speech which created the worlds". It is the subtle aspect of speech that transforms itself into the forms of the worlds.

śrutyantaramapi "vācaiva viṣvaṁ bahurūpaṁ nibaddhaṁ
tadētadēkaṁ pravibhajyōpabhunktē iti"
– Tattvavimarśinī 05-06:c

In another śruti it is said: "The world with its multiple forms is composed of speech only, and having divided, one part of it is being experienced."

Vaikhari is the culmination of the process of how sound vibrations create the universe through the voice or Word of the Creator, that has now moved through all four modes of sound, *and* our human voice. Whilst there is a differentiation between vaikhari and madhyama, they are a mirrored process: as above, so below.

The more we own our words and voices as creators, the more we come to Realize this whole process is happening within ourselves. We can shape and create our world in harmony with the modes of sound that come from the original Word of creation.

Our speech is an echo of the Word, an echo in human form of Source expressing itself. As such, our speech is an instrument for transmitting pure consciousness. Our voice is the most powerful tool we have to create, and is the first instrument that gave rise to all music and to all instruments.

The Spirit of words known as *Kotodama* in Japan and *Vak* in Sanskrit has a powerful effect on our realities. Words can hurt, inspire, enlighten and destroy. They affect all we do, forming how we think, and what we think, for words are the foundation of our thoughts, our cultures, our media and the bedrock of society itself.

The subliminal messages delivered by words are impactful, for words create thoughts, thoughts create ideas and ideas shape the world we live in.

Words form the thread on which we string our experiences, as Aldous Huxley said. Words are spells. Thoughts are spells. Your intentions are spells. Your inner conversations are spells. Your beliefs are spells. Poems are spells, mantras are spells, and hymns are spells. Your texts, your emails, your conversations with others are spells. You mould, shape and create your world with every spell.

Words seduce, beguile, intrigue and transport us. They create worlds and agreements we are not even consciously aware of. Words spin a web within which we live out our lives, liberating us or imprisoning us. Each word carries an energetic frequency that people react to in unconscious ways based on their associations with it, often from childhood.

For example, the word "God" evokes a diverse range of reactions from fear to joy. Words give form to our beliefs and our conditionings, and then guide our expressions, actions and emotional states.

In Sanskrit, words are understood to weave the matrix of space-time itself. This matrix is constructed of letters and words, of currents of sound and mantra, that interconnect all life through streams of sound vibration. For example, upon his enlightenment, the Buddha saw a web of Sanskrit letters glowing in a golden tree stretching from heaven to earth. Similarly, yogis use sound and word meditations in different chakras to align to and eventually transcend the matrix of space-time.

In the West, Kabbalists use word-sound meditations based on Hebrew, where each letter has a number. These letters have been shown by Stan Tenen to arise out of the *torus* or quantum spiral that forms the field and wave of creation, and indeed these letters form the torus spiral itself.[48]

Modern philosophers such as Terence McKenna[49] affirm this, saying that "the syntactical nature of reality, the real secret of magic, is that the world is made of words. And if you know the words that the world is made of, you can make of it whatever you wish."

Sound, word and meaning are one in the sacred languages, unlike languages like English where meaning is derived from association, i.e. what we have been taught a word means. For example, you can expand your consciousness with Sanskrit without having to learn its meanings if you *feel* the vibration of it, as each word is a holographic pulse, transmitting its principle in a single moment.

One word in Sanskrit can take paragraphs to translate into English, as its essence is multidimensional, not limited to the third dimension alone. You deeply and experientially feel these sounds and words as they bypass the middle man of the rational and conditioned mind, resonating into the deep structures of the brain and nervous system. This is where the more multidimensional modes of sound come into play, expanding one's consciousness beyond our everyday speech and words.

This is why Sanskrit is used so extensively in spiritual practices such as mantra and *kirtan*, as it very effectively uplifts your state of consciousness quickly, as hundreds of millions of people worldwide experience every day. We directly resonate with sacred languages such as Sanskrit because they are the roots of most present-day languages. They resonate with us on a primordial feeling level beyond the logical, conditioned, what we have learned, memory bound self.

Vak

In India the power of speech is known as *Vak*. Vak is associated with Saraswati, Goddess of speech, the Word, wisdom and music. Speech is associated with the fire element, which brings about form, and catalyses, ignites, warms, dissolves, manifests and brings new energy, power and direction.

The science of *Vak* is found in seed syllables, or *bijas*. Seed syllables are the simplest and most universal sounds. From *bijas* come mantras, more complex forms of sound patterning. This can be likened to the beginning of the universe, where the simplest and most powerful sounds issued forth, such as the vowel sounds common to all languages in *AIUn*.

As creation diversified, creating more complex and manifested forms, mantras of seed syllables, vowels and consonants became the sound signatures of these harmonic clusters. Consonants act as links to form matter˙ (vowels are seen as pure light), and were inserted between vowels in order to create more complex forms.

The essence of sound is in its elegant simplicity. *Bijas* are the beginning of speech, from which all speech arises from, and are the most potent ways to access the universal sound current.[50] Universal seed syllables are gateways into the universal sound current and resonate with every chakra and the whole nervous system.

These sounds can resonate, penetrate and excavate much of what lies in the way of each chakra blooming open. The most well-known of these seed syllables is AUM. Whilst there are many seed syllables for each chakra in different traditions, these work on certain levels. When one is clear and sensitized enough, one can use universal seed syllables effectively to connect with the fundamental elements of creation and for personal transformation.

RLRK: Maya and The One

अकारो ज्ञप्तमात्रं स्यादकिारश्चतिकला मता ।
उकारो वषिणुरतियाहुर्व्यापकत्वान्महेश्वर: ॥९॥

Rḷk' sarvēśvarō māyāṁ manōvṛttimadarśayat
tāmēva vrittimāśritya jagadrūpamnajījanat
9

RLRK is the One awareness
viewing and birthing through Maya
all the infinitesimal movements of mind
operating in every being.

These quantum streams whirl, expand and spin,
modifying the indivisible field,
moulding perceptions and parameters
creating the blueprint of movables and immovables,
for the ever-moving forms of the universe.

ऋळक् सर्वेश्वरो मायां मनोवृत्तमिदर्शयत् ।
तामेव व्यक्तमिाश्रतिय जगद्रूपमजीजनत् ॥१०॥

Vrttivrttimatoratra Bhedaleso na vidya
Candracandrikayoryadvat Yatha vagarthayorapi
10

Without knowing any difference between them
and the field they are in,
infinitesimal movements cascade in and through each other,
modifying the indivisible field
diverse in their unity, many in the One.

There is no difference between the moon and its rays,
no difference between a word and the power of its meaning,
no difference between these modifications
and the One who creates-sustains these modifications.

वृत्तविृत्तमितोरत्र भेदलेशो न विद्यते ।
चन्द्रचन्द्रकियोर्यद्वत् तथा वागर्थयोरपि॥ ११ ॥

svecchayaya svasyacicchaktau visvam unmilayat yasau
varnanam madhyamam klibam Rlrvarndvayam viduh

11

The Self multiplies Itself, unfolding a shadow of Itself,
forming the faculty that perceives individuality,
which is present in every being.

Through the will and desire of the One,
the whole range and spectrum of dimensions
of the universe unfold.
Apparent boundaries invisibly form,
in order to perceive the universe in many different ways.

The wise Know the pair of syllables
in the middle of *Rlrk* are neutral;
R is the One awareness, and K is the created
R is the Knower, and K the known.

The trinity *A I Un* is the Clear Light of Awareness, the Action of Sakti and Awareness-in-Creation Visnu in one quantum field. The One duplicates Itself into The One, 1+1, and the other One into Two. The other One marries the Two, 2+1, and they become Three in *A-I-Un*. This of course is the Fibonacci sequence, which was originally discovered in India at least 700 years before Fibonacci and is rightfully called the *Pingala-Hemachandra* sequence.

Then arises *RLRK*. *RLRK* is a reflection of *A I Un*, the first reflection of the One to Itself within Itself. There is now something to know for the Knower. In Rlrk, Awareness goes back within Itself. As *AlUn* unfolds Its own nature in *RLRK*, It begins to think that the bliss and consciousness It is will lessen. As this happens, it begins to take Itself back into Itself in *RLRK*.

In RLRK, the outwards expansion has paused, yet something is still happening in the process. In Kashmiri Saivism, it is said that this stage is where Shiva has not yet accepted the existence of the universe in his own nature yet.

What has started to expand outwards to form the quantum field in *AlUn* now comes back into Itself to rest in Itself. AH or Shiva now resides solely in its own consciousness-bliss. It stops to be Singular, to rest in Itself.

When we look at the letters of RLRK, the first short letter *RI* is the intention to return to its own nature. Then there is the confirmation of this intention in the long letter *RI*. The next movement in the short letter *LRI* is the establishing of the Will to return to inside Itself, with the 4th long letter *LRI* confirming this establishment and carrying out the movements to come to rest in Itself.[51]

The great sage grammarian of Sanskrit *Panini* called these four letters the "eunuch state" referring to the Sutra where it is said that *"the pair of syllables in the middle of Rlrk are neutral."*

These four letters are known as *amrita bija* or bliss filled letters, because they reside in One's own nature, which is bliss.

The whole process of universal expansion in the indivisible quantum field *AIUn* is now carried back into One's own nature through RLRK and its *Iccha Sakti* – the energy of Will. From this eunuch state of *RLRK*, creation comes forth again refreshed in the next class of sound *EOn*.

In another translation of the letters *RLRK*, *R* is awareness, from which arises its consciousness Maya *L*, which reflects awareness to create a sense of individuality. These infinitesimal quantum movements form the faculty that perceives a sense of individuality (*visvam*) that is present in every being, as well as reaffirming the Great "I" of Shiva.

The moving reflections and infinitesimal movements of this dance is *K*.[52] Maya *L* is totally enfolded in *R*, encircled by *R* and held by *R*. *R*-awareness unfolds its consciousness within Itself in a moving reflection *L* that is the first ever experience, experiencing, and forming of the Knower of all experience that is the Great I Am, *and of all sentient beings*.

Mano-Vritti

A I Un unfolds Itself through this "mirror mind": the projection power of Maya or *L*. RLRK manifests through infinitesimal quantum light particles, known as *manovrittis*. These manovrittis flash in and out of existence, forming the idea of experience itself *and* an individual experiencer. Before this point in creation, there is no experience or experiencer.

In quantum physics, this Dance of quantum particles creates the movements between 0 and 1, flashing in and out of existence in a rapid succession of instantaneous,[53] infinitesimal movements fluctuating within the quantum field.

> *The world is movement, a rapid and continuous rapid succession of flashes of energy. There are no objects "in movement," it is movement which constitutes the objects which appear to us, and all objects, all phenomena of whatever kind, are constituted by a rapid succession of instantaneous events.*[54]

These quantum streams are bliss in motion whirling the synchrony of all life, expressing all its myriad movements in ecstatic waves, creating, recombining and multiplying in virtually innumerable fractals, only to dissolve again.

These reflected movements of the manovrittis form an experiencer *and* thus the potentials of all experience through the "I" or Knower consciousness. The Knower and the known manifests for the first time.

RLRK is awareness and its reflection revealing a sense of individuality. Maya is this mirror movement, a mirror that only sees itself, appearing when awareness desires to experience itself. Thus, Maya is the consciousness pattern for experiencing as an experiencer. Without Maya, individual being and the possibility of experiencing would not exist.

Maya uses the energy of Sakti *I* in a set of given parameters, boundaries and geometric structures in order to reflect Itself. For energy to be in motion, it must be limited in some way within a pattern. This manifests in RLRK through the *manovrittis*, infinitesimal movements of reflection and quantum particles.

Nobel Prize winning scientist Roger Penrose states that quantum particles are like tiny curvatures in space-time geometry (as Einstein's Theory of Relativity had done for large objects like the sun). Superposition states of multiple possibilities, or delocalized particles, could then be viewed as opposing curvatures, and hence "separation" processes[55] in the fine scale structure of the universe and its space-time geometry. This is what manovrittis are.

The manovrittis are "proto-conscious moments" which have no memory, no meaning and no dualistic context, but have phenomenal "qualia" – a precursor to the capacity to experience. They may be like the notes and sounds of an orchestra tuning up before they play their symphony of experience.

Hameroff and Penrose found these quantum vibrations in the *microtubules* in brain neurons were "orchestrated", leading to their idea, echoing the Maheshwara Sutra, that consciousness is like music in the structure of space-time.

In Sanskrit, *Mano* means the primordial mind, the first experiencer experiencing something, moving fluidly between the newly created *this and that* or all possible values and numbers between 0 and 1. Its infinitesimal fluctuations of quantum particles give rise *unmilayat* to the first perception, the first experiencer and the new notion of actually experiencing.

Vritti is a whirling vortex of spiralling activity, turning around and around. As Patanjali shares, *Vrittis* are any piece of content and any experience, including all thoughts, dreams, perceptions, imagination and states of consciousness.

There are five main *vrittis*: correct knowledge or *pramana*, which consists of positive thoughts leading to health, mental clarity and expansion; incorrect knowledge *viparyaya* or negative thoughts leading to destruction, bondage and ignorance; *vikalpa* or dreams, imagination, fantasies, mental constructs, abstractions, conceptualizations, experiences and oscillations of the mind; sleep or *nidra* and finally memory or *smrti*.

In our human experience, vrittis arise from our actions and experiences, which leave imprints on the clear mirror of RLRK. These then become repeated patterns and behaviours that make each one of us "unique". Our desires and repulsions, our choices, preferences and dispositions, are put together in a stream of moving thoughts, *the moviemaking power of Maya*, to create a sense of individuality, *visvam*.

Vrittis form an ever-moving thinking and narrating commentary, labelling and categorizing experience from the perspective of an individual experiencer. This mind interprets, comments on and defines everything it comes into contact with, and gradually becomes identified with these experiences.

This apparently solid construct of "I" formed by this ever-moving stream then comes to assert and believe that experience is all there is, and that *it is* these experiences. It defines all experiences as either good or bad, *vrittivritti*, and evokes emotions and mental tendencies to push away or invite more of these experiences.

When we slow down this mind, like a movie slowed down frame by frame, or when we zoom in on an individual pixel on a screen, we see that who we believe we are, who we perceive ourselves to be, is a sequence of these vritti movements or visual sounds.

These visual sound movements create your own personal movie narrative, with Maya as the director, forming a sense of individual self – based on a rapid succession of infinitesimal movements composed of hundreds of different voices, pictures, patterns, memories and associations. These all create the narrative story of your I, woven together through these mental movements into one continuous flow.

This Maya is twofold: it enables Shiva awareness and its consciousness Sakti to experience and enjoy Itself in many different ways. It also enables each one of us, as part of Shiva and Sakti, to have experiences which we too can enjoy, but which we usually become enmeshed in *if* we are not situated in pure awareness *R*.

If one gets enamoured by and identifies with the manovrittis of movement and experience, one becomes engrossed in worldly appearances and the need for more and more experiences to fulfil oneself and to make one happy. "Maya is the power of a storyteller to weave threads of characters together into an absorbing tale that appears real, and which makes us forget everything else."[56]

In *Rlrk*, consciousness and the possibility for any and all experience as an experiencer arises from awareness, which is free from any identification with experience or desire. Awareness *R* stands independent of experience, reflection and the vrittis, yet can still enjoy them. Awareness is this blank screen, without images; consciousness is the images and movie playing on this screen.

"A screen is intimately one with all images, and at the same time free of them; so our true nature of luminous, empty Knowing is one with all experiences, and yet at the same time inherently free of them."[57]

RLRK in the Foetus

Mano is the breath of primordial mind that enters the foetus at the beginning of life, establishing the RLRK faculties in the foetus as it expands into five cells. *Mano* is also the breath that leaves the body at death, rendering it 21 grams lighter.

In the *Vigyan Bhairav Tantra* v. 21, where Shiva teaches Shakti 112 meditations to bring her into Self Realization, there is a practice which beautifully describes the manovrittis:

> *Focus on your tailbone, vibrating with luminous space. Follow the flow of white light in the middle sushumna channel of your spine, flowing in emptiness. Here, where particles flash in and out of existence, is the origin of mind.*

Curiously, when conception happens in the womb between a female egg and male sperm, infinitesimal flashes and sparks of light have been filmed. Billions of tiny zinc atoms are released at the moment when the female egg is pierced by the sperm, and these "fireworks" last for about two hours.

MAYA

Maya does not mean illusion in the same sense in which the English word is used. "Maya" means to measure. You cannot measure anything unless you have a point. So, if the centre is absent, there is no circumference at all... so, in that sense, anything you experience based on knowledge is an illusion.
– UG Krishnamurti: Throwing Away the Crutches

Maya sakti in its reflective energy forms the faculty of individuality (*visvam*) that is present in every being. When this energy is known in pure awareness R it is called *svatantriya sakti*, pure energy Will of the One that weaves creation. When you perceive this svatantriya sakti through your own "I" filters and conditioning, it becomes Maya Sakti.[58]

This belief in the experience of Maya Sakti binds you by making you believe that the universe is separate from your own nature. This is its illusion. When you realize Maya Sakti for what it actually is, it becomes Svatantriya Sakti. It is just from where your own filters and observation viewpoint make you see it.

Another way of seeing this is to understand that the purpose of Maya is to help *A*/Shiva/you, recognise Shiva's/your own true nature. Shiva/you does not recognise Shiva's/your own nature in *A* because Shiva/you is already there and is all there Is.

But Shiva/you wants Shiva's/your own nature to be recognised and enjoyed. And yet, because Shiva/you is already there in *A*, there is nothing to recognise. Therefore, in order to recognise Shiva's/your own nature, Shiva/you must first become ignorant of Shiva's/your own nature. Only then can Shiva/you recognise Self.

The whole universe is the means by which we can come to recognise ourselves as Shiva. This is the purpose of the universe, the how and why it was created. This is the intention and desire of *A*. Maya is the instrument of this.

In the Sutra, the clear moon is likened to the still mind of awareness, with the rays of the moon clearly reflecting its source. But with even a small ripple or vritti, the moon's appearance is distorted: the mirror is smudged and the reflection unclear. These ripples of the vrittis in our human experience partially arise from memory.

It is impossible to live as one person with an individual history, or to possess our being in a continuous fashion, without the memory threads that link past and present to the idea of the future.[59]

Memory and its stories define who we think we are, and who others are in our own minds. Memory, whether it is "real" or imagined subjectively through our own filters, shapes our mind and emotions, conditioning our reactions to whatever is presented to us in life.

Memory makes it possible for us to live as an individual, believing we are separate from others. This has a purpose and usefulness, as I could not even lift a glass or type a word on a computer without having a memory of how to do so, and without having a sense of separate identity to the computer or glass. Yet, when this sense of separation and memory becomes accentuated and out of balance, it leads to forgetfulness, division and conflict.

Movement and memory are bound together by the forces of time and the movie director Maya, which constantly seeks out and then interprets experiences. Your personal movie of reality behaves according to these filters and mental modifications of the imaginal dream maker and moviemaking power of Maya: the director of the vrittis.

This all stops you from experiencing directly WHAT IS. In the present moment, one no longer experiences self, others or the world by past impressions; everyone and everything is seen as it truly is in the present moment of *Un*, unfiltered by the vrittis.

In RLRK, the seeds for memory, movement and time are laid, all of which are necessary to create form. Time is necessary as it measures energy in motion through which evolution, or measurements of change, can happen.

Time defines the existence of forms, the body and the mind; without time, there are no forms, no names or vibrations, no objects or nouns, no life and no death, and no way to measure, unfold or compare anything.

Time is the measuring rod for all life to be born, move and unfold their forms, sustain themselves, and eventually die. Time is the measuring rod of life and the sense of individuality: of a separate self-existing. Time is Maya's instrument.

Vrttivrttimatoratra Bhedaleso na vidya
Candracandrikayoryadvat Yatha vagarthayorapi.
Without knowing any difference between them and the field they are in, infinitesimal movements cascade in and out of each other, diverse in their unity, many in the One. There is no difference between the moon and its rays, no difference between a word and the power of its meaning.

In Sanskrit, when the same word is repeated twice after each other, it can indicate polarities and/or the opposite of the first word's meaning. In the quantum field there is no discrimination or separation in what is occurring in the same unified field, yet Maya is now creating subtle movements, differentials, measurements and modifications in her reflecting of awareness.

There appears to be a separation between consciousness and awareness with the vrittis, which flash in and out of existence, but there is not. This is beautifully shared in the Sutra, where the moon and its rays are the same, just as the sun and its rays are the same, just as a drop of water in the ocean is the same as the ocean itself.

The moon and its rays/reflections are from the same source, conjoined in eternal yoga. The moon is cool, calm, serene, clear: its reflections are held in it, the cause. All effects contain their cause, just as the cause contains in potential all its effects. The moon is clear, still awareness reflecting itself through its rays in order to See itself.

Sanskrit sounds and their meanings are united through *the power of vritti*, the power of a word that expresses its meaning. In Sanskrit, sounds and meanings are indivisible, felt as one burst of sonic information consciousness. The vibration of the word reflects the meaning, and each contains the other, each speaks of the other and cannot be dis-associated.[60] When you sound Sanskrit words, you can feel their meaning as the sound and its form as one. What it sounds like is what it does.

In Nada Yoga or Sound Yoga, this is summed up beautifully in one saying (as we saw earlier in the sutra): *Sravanam, mananam, nididhysanam*: Hear, Reflect, Realize. When we Hear and experience openly and innocently, without judgment, vrittis, conditioning or preconception, then we reflect the essence of what is being experienced, and Realize what it really is.

Awareness witnesses all movements, vibrations and modifications without being involved in it, without any limits, and yet is part of this unfolding as it Sees and therefore brings into existence what it Sees. Maya is the subtlest yet most intrinsic consciousness architect of reflection, the fundamental and first shape making power of appearance, a process of becoming emanating sublime fluctuations from within awareness.

Svecchayaya svasyacicchaktau visvam unmilayat yasau
Varnanam madhyamam klibam Rlrvarndvayam viduh.

The Self multiplies Itself, unfolding a shadow of Itself, forming the faculty that perceives individuality, which is present in every being. Through will and desire the universe arises; apparent boundaries form in order to perceive the universe in many different ways. The wise Know the pair of syllables in the middle of Rlrk are neutral; R and K are Creator and created, Knower and the known.

Consciousness reflects and therefore multiples Itself within Itself in RLRK. This movement creates the first sense of individuality, which becomes a faculty present in every sentient being as they form inside the womb.

With Shiva's Will returning into Itself with Sakti-desire and Visnu as the all-pervading sense of sentient existence, Maya forms as *L*: the mirror of awareness. This consciousness now reflects into virtually infinite fractals of Itself, able to look at itself and see itself in many different, new and exciting ways.

The wise know the pair of syllables in the middle of Rlrk are neutral. R and K are Creator and Created.

"The pair of syllables" refers to the eunuch state Panini explained earlier as *amrita bija* or bliss filled letters, because they reside in One's own nature, which is bliss. This return into One's own inner nature through RLRK is through *Iccha Sakti* – the energy of Will.

Awareness *R* is The Knower in which all that can be known (i.e. all forms, processes, experiences, experiencing and the experiencer) appears, and is also That out of which all this arises. "Me" and "the world" *K* are the known and created, moving, changing, appearing and disappearing in That Awareness which never appears or disappears. *"Look in all experience for That which does not appear, move, change, evolve or disappear, and know yourself as That."*[61]

Sounding RLRK

RLRK is awareness birthing the architect of consciousness: the Knower appears with the known. RLRK gives certain rules and stability for movement and perception. **R** is the vowel governing the cerebral class of sounds. This soft "r", pronounced "true", is pronounced with the tip of the tongue turned towards the roof of the mouth, just short of the upper palate.

L is the sound of Maya. The tongue tip points to the lower edge of the upper dentures. In **LR** the tongue points to the lower edge of the upper teeth.

Note the difference between the two **R** sounds. The first R has the tip of the tongue turned towards the roof of the mouth, just short of the upper palate. The second **R** in **LR** has the tongue pointing to the lower edge of the upper teeth.

The subtle movement between the two has the tongue curling, unfolding, then re-curling in **K** to form RLRK. When you sound RLRK correctly, the tongue movements help to still the mind. Preparatory Kriya Yoga techniques help do this too.[62]

Sound **RLRK** focused in the higher third eye, a few centimetres above the point between the eyebrows usually associated with the pineal, and vibrate the upper palate. Emphasise each sound gently, resounding these spaces with each syllable. Try doing it slowly, then sound it fast, for at least 50 times.

In *Vigyan Bhairav Tantra* 11, where Shiva teaches Shakti 112 meditations about Tantric Self Realization, he says:

Inside the skull there is a place where the essences of creation are consecrated and mingle: the light of awareness and the awareness of that light. The divine feminine and divine masculine here illumines all space. Rest in this ever-present light that is already here, and everywhere.

EOn: Cosmic Egg

एओङ् मायेश्वरात्मैक्यवज्ञानम् सर्ववस्तुषु।
साक्षत्विात्सर्वभूतानां स एक इति निश्चितम् ॥ १३ ॥

ēōṅ māyēśvarātmaikya vijnānaṁ sarvavastuṣu
sākśhitvātsarvabhūtānāṁ sa ēka iti niśchitaṁ

13

EOn is Maya in union with Isvara.
Maya reflects the Lord of creation Isvara
as the consciousness of eternal essence,
within the consciousness of eternal essence,
from which all things arise.

The witness of everything is here as One:
there is nothing apart from Its own existence.
This is for sure!

Awareness witnesses and therefore brings into being *all potential qualities* of creation through *Māyēśvarātmaikya*, Iśvara and Māyā in union. Isvara is the organizing awareness of the formless in creation and is the Lord of Creation, unbound by the forms, appearances and limits of creation. As Ramana Maharshi shares in the book *Conscious Immortality*, "Ishwara has individuality in mind and body, which are perishable, but at the same time He has also the transcendental consciousness and liberation inwardly. Ishwara is immanent in every person and every material object throughout the universe."

This union of Ishwara with Maya in *Māyēśvarātmaikya* enables the continuation of creation after the inward movement of *RLRK*, which established the individual Great I AM *and* the faculty of individuality in every being. The mirroring of *Māyēśvarātmaikya* of the limitless Ishwara and the power of limits Maya now refracts this movement within Itself to reflect and replicate Itself, thus Seeing Itself, in order to form the Cosmic Egg.

In *EOn* there is the first ever experience of unity, *whereas before this there is not even a sense of unity to experience,* just the One moving within the One as Itself. This unity can only be experienced and Seen through the witnessing consciousness, which is now created.

This occurs in the invisible quantum field of energy and information existing before and beyond space and time. Nothing physical or material exists here, and it is beyond anything you can perceive with your senses.[63] This unified field of quantum information in the Cosmic Egg of *EOn* governs all subsequent laws and classes of sound.

The next multihued word in the Sutra is *Vijnanam*. The sage Sri Ramakrishna defines *vijñānam* as the capacity "to know God distinctly by realizing His existence through an intuitive experience, and to speak to Him intimately." This intuitive, direct, immediately spontaneous insight and perception of clarity is birthed in *EOn*, and a person who is a *vijnani* is a Gnostic, one who has direct insights, experiences and knowings of reality that transforms them.

This Gnosis or *vijnanam* is how sages realized and brought forth the Vedas, Kabbalah and Buddhist and Egyptian wisdom teachings. This Gnosis, Seeing or Lightning Flash of direct realization[64] arises from *vastusu*, an eternal and imperishable substance from which arises all created things.

As *vijnanam* arises from vastusu, it only Sees vastusu, and this is why those who enter the infinite bring forth the wisdom of the infinite. Only the infinite knows the infinite, and nothing else, just as the finite mind can only know the finite, and nothing else.

Vijñana is consciousness, ever present in all experiences from your point of view. It is "your" consciousness, the part of "you" that is always there, ever present, independent of any experience, relationship or situation.

The Witness

The ability to observe without evaluating is the highest form of intelligence.

– J. Krishnamurti

The reflected consciousness of *Maya* is witnessed *sākśhit* as *Māyēśvarātmaikya*. Creator Sees Itself as Created within Itself. The Knower knows itself in this unfolding. "The universe as we know it is the joint product of the observer and the observed," as Teilhard de Chardin shares, and this is what happens in the union of Maya and Ishwara in *Māyēśvarātmaikya*.

Maya first arises in the second movement of creation *RLRK*, and is the second last aspect of consciousness to be fully cognised before a human becomes fully aware. This cognition occurs through the singular witness, That which is part of every experience, yet free, independent and detached from every experience at the same time. Only witness is definitely, *nischitam*, omnipresent.

Since It is omnipresent, the witnessing awareness is one and the same in each individual, because each individual arises from This. This one and the same faculty of the witness in each sentient being echoes the previous verse RLRK, where Maya helps birth the faculty of individuation in every sentient being.

In *EOn*, this singular witness awareness sees *all potential qualities* of creation in Itself, and therefore brings this forth in seed or egg form, witnessing Itself from within Itself in the (Maya) screen of created and the creator.

Without the witnessing consciousness time and space cannot be recognised. Yet the consciousness that witnesses, and therefore creates, time and space transcends time and space. In *EOn*, awareness witnesses and therefore conceives of anything and everything ever possible, including all timelines, all possible futures and all conceivable states of existence.

According to quantum mechanics, matter and form do not exist when you are not looking at it, so the universe does not exist if no one observes it, as observing something brings it into being.[65]

Witness does not emphasise one over another, or judge things as good or bad, comparing or contrasting. In witness, there are no enlightened ones or ignorant ones. Witnessing is awareness of the whole moving playing field of consciousness and all its contents. Your thoughts and emotions, words and actions, are a part of this, but the witness is not affected.

Witness sees that who you think you are is a set of vrittis, movements of infinitesimal energy and mental modifications creating different forms that arise and disappear so quickly in the moviemaking power of Maya that in the waking state you cannot cognise it.

Everything and everyone in your life is a reflection and expression of these *vrittis*, which are your own reflections and projections.[66] The vrittis are also composed of your emotional wounds, traumas and patterns, with its attendants of rationalizing, excusing and denying your emotions, known as *spiritual bypassing*. People may believe they are in the witness when in fact they are still in the vrittis. How does one know the difference?

The witness does not shut off, isolate or alienate itself from feeling anything or anyone. It experiences everything whilst being That which is beyond any experience, thought or emotion. It is in the world but not of it, present everywhere. Witness is a natural state untouchable by any event, situation, person or emotion. Free from reaction and response, *whilst at the same time feeling and flowing through* it all.

Everything you observe occurring in your life and within you is neutral, until you make a decision as to what it *should* mean for you, based on memory, conditioning and the vrittis. Witnessing is Presence, *rather than what you think, emotionally feel or imagine is present.*[67] Witness observes the movements, actions, thoughts and feelings of the vrittis without involvement, yet is not separate from this dynamic, ever-flowing ocean of movement.

Witness Sees the dream of the vrittis that arise from Maya's mental power and reflections. Witness knows it cannot be harmed. No matter what the content is, witness is not affected by it. Peace is its hallmark, occurring when our mental and emotional energy is sublimely liberated into a quality of presence in the midst of all activity.

No longer is there a dependency on the external, and no anguish arises at what comes and goes. There is no longer a sense of striving *or need to get.* Blame and praise are felt as equal and the fear of death, of the body-mind perishing, does not exist here. This does not mean you are cold, unfeeling, aloof, rigid and lacking in dynamic life-force, human engagement and joy. On the contrary:

"Once you are in the witness you find that you love what you see, whatever may be its nature. This choiceless love is the touchstone of awareness."[68] If this love is not there, you are not fully in the witness.

As Patanjali describes, *"patterns of consciousness are always known by pure awareness, their ultimate unchanging witness. Consciousness is seen not by its own light, but by awareness."*

Quantum physics tells us that everything that is observed is affected by the observer. The behaviour of particles is inextricably linked to the presence of an observer. Every individual under the influence of Maya sees a different reality because everyone is creating/perceiving what they see through the individual lens created by Maya, the faculty of individuality she reflects. What we perceive is a process that involves this consciousness.

Maya consciousness does not have its own light, as it is reflected from awareness. Awareness is living light, as it is not reflected off anything else.[69] Maya consciousness cannot see itself, any more than a TV can watch itself, even though it can display a vast number of different channels and programs, each one offering an apparently realistic view of life.

Once you become aware that there is "something else" watching what "you" are reading right now, that there is a watching awareness present behind your vision, actions and perceptions right now, doing nothing, being nothing, then you can access witness consciousness or the watcher.

In the witness, the first act of perceiving happens: the first looking at Self[70] within Self. This is the beginning of unity in diversity, the many in the One, both states simultaneously occurring. As Christ said, *"When you make the two into one, when you make the inner as the outer, the upper as the lower, when you make male and female into a single one, so that the male shall not be male, and the female shall not be female, then you will enter the kingdom."*[71]

In *EOn* all possibilities and potentials are seen and created, *including* the possibility for the finite, *and* the possibility for the finite to become infinite. As the possibility of an individual is seen here, it also becomes possible for an individual to become One with the One here, in a sheer and potent intensity of inclusiveness.

In the cognised maya shakti of *svatantriya shakti*, everywhere you turn you see a mirror image of yourself. Wherever you look, there you are. You are in everything, and all you can see is you. This consciousness, the reflection of awareness in movement, is the dance of Isvara and Maya.

Just as the moon is not separate from its rays of light, and just as a word is not separate from its meaning, so is the dance of Maya consciousness unfolding from her creator awareness.

Maya arises from awareness as its consciousness, bringing forth the created from within the creator in the same substance, unfolding Itself.

In this mirror dance arises all the potential possibilities of time, timelines and being able to have experiences in time. Maya is the experiencer and the powers of experience, the architect and primordial imaginal dream maker.

All that an experiencer can experience is Maya. Without her, there would be no experience, experiencer or even the possibility of experience. This is *A*, pure awareness, where *"our true nature of Knowing is one with all experiences, yet at the same time, inherently free of them."*[72]

Witnessing is to See as God Sees. This is established in *EOn* as a fractal hologram of awareness and its play of reflections. In the two classes of sound *RLRK* and *EOn*, Maya is named as:

- A form of consciousness reflecting awareness.
- A tool and instrument for Self-Recognition or self-delusion.
- The perception of individuality in every sentient being: the moviemaking power of "I".
- The architect of creation and its measures, which include:
  ◊ all the possibilities and potentials of time and timelines
  ◊ all possible experiences, the faculty of experiencing and the creation of the experiencer itself
  ◊ memory and its eight forms (shared later in the book)
  ◊ the five vrittis or movements of mind
  ◊ the five constituents of *vijnanam* or consciousness that forms perceptions, wisdom and knowledge (shared later in the book)
  ◊ the form reflecting power that results from all of this combined.

- Maya's tools are also the three gunas of light, action and stability, as Krishna mentions in the *Bhagavad Gita*. These three gunas are explored in the 13th Class of Sound later in the book.

Curiously, there is a remarkable similarity between the creational energies of *Maya-Ishwara* or *māyēśvarā* and the Egyptian creational pair of deities known as *Seshat-Thoth*. Maya is the great measurer, who works in union with Ishwara to measure out perceptual boundaries in the quantum field in order to create form. The Egyptian "goddess" Seshat does the same thing in union with the "god" Thoth, the architect of creation from the quantum field.

As author Ani Williams shares,

before creating the foundation of an Egyptian temple, there was a ritual called *Stretching the Cord*. Their measurements were accompanied by incantations and established the harmonious position and proportion of the temple on earth in accordance with the heavens and the cycles of time.

The goddess Seshat chanted specific sound formulas as she stretched the cord of sacred measure. The tones are harmonious and perfect, as are the calculations for the acoustical design of a new temple. Seshat is crowned with the seven-pointed star, the number of spirit and the cycles of time. Thoth brings the ancient knowledge to guide the foundation ritual.

The temple mentioned is of course the universe, the human form *and* a physical temple, with all of them composed of sound vibration. The same laws applied to the creation of all three, hence the great saying or *mahavakya*, "As above, so below."

As we shall see throughout the Maheshwara Sutra, there are remarkable similarities between the ancient Vedic science of consciousness-creation and the ancient Egyptian science of consciousness-creation. This has been reflected in modern times in esoteric Christianity, much of which is copied and descended in diluted form from ancient Egyptian wisdom.

The Cosmic Egg – Hiranyagarbha

To pitch darkness they go who worship the Unmanifested. To a greater darkness than this go those who are devoted to the Manifested Hiranyagarbha. He who knows both the Unmanifested and the destructible together transcends death by revering the destructible, and attains immortality by revering the Unmanifested.
– Isavasya Upanishad

Sanskrit is a multidimensional vibrational matrix that conveys ineffable truths. This can only be Realized by immersing in the vibrational consciousness of the sound itself. Each word-sound is a hologram containing multiple layers of information, knowledge, wisdom and Insights (*vijnanam*) that evoke and express states of consciousness.

Each word-sound needs to be resonated aloud and internally, reflected upon, deeply listened to, moved into differing chakras and centres of consciousness, energetically contemplated, and therefore Realized.

This is one of the great sayings of Nada Yoga: *sravanam mananam nididhysanam: Hear Reflect Realize.* This is how one realizes the sound-consciousness essence of any form, and in so doing this process changes you, shifting your experiential and intellectual perceptions, using and uniting both sides of the brain.

One has to become the consciousness of the sound in order to understand and unpack all its nested levels. True gnosis of a Sanskrit verse or word flows like a wave, changing the translator internally, bringing the translator into that vibrating consciousness, altering their perceptions into this essence.

In this verse, the next word, which is most delightful, is *sarvavastusu. Sarva* means everything, all. *Vastusu* means dawning light appearing from eternal imperishable essence substance, or quantum matter. This imperishable quantum substance forms *Hiranyagarbha* or the Cosmic Egg in *EOn*, the quantum container-womb that incubates and gestates all the holographic blueprints of all forms and polarities, to then later birth the universe, ourselves and all worlds from these templates.[73]

Hiranyagarbha is the culmination of the previous 2 Classes of Sound or Steps of Creation AIUN RLRK into an Egg of Self Contained Infinite Potential that is EOn. *Hiranyagarbha* is seen in the Vedic vision as a radiant egg floating in emptiness, emerging on an ocean of ether from which creation comes forth in a luminous fire mist (hydrogen gas). This Egg of imperishable substance is "the source of creation on all planes, and is in essence eternal, enduring during the great cycles of time."[74]

Within the Self-containment of The Egg[75] the initial sets of quantum polarities unfold within themselves to form all the potential possibilities of creation. Awareness witnesses and therefore guides and forms this dawning light.

This Egg of Self Contained Infinite Potential is the culmination of all previous movements as they settle into an encoding womb matrix made of quantum matter. The Egg is a Self-contained unified circuit cycling quantum infinitesimal movements in quantum matter. As the extreme of the contained circuit is reached, it switches direction/movement/perspective/perception in order to impregnate itself, creating another Self-contained potential in this endlessly connected circuit or infinite battery.

This infinite battery is where the full range of creation's potentials and possibilities are formed, in a series of quantum blueprints.[76] It is this point in creation that we, as humanity, need to access in order to create like gods, to create a quantum civilisation.

In the fluidity in unity of EOn, the quantum holographic blueprints of masculine and feminine principles are fully formed before the Big Bang, before matter and light as we know it. The birthing of physical creation from the Big Bang is preceded by the Cosmic Egg of EOn, from which physical creation manifests into matter in the toroidal wave-field of the universe in the next Class of Sound *AIAUch*.

In contemplating Nature's being, Know the One as many seeing. In and outer coinciding, nothing in or from our dividing. Open secret revelation! Grasp it without hesitation. Free of seeming Truth's confusion, Revel in the serious game. Separateness is the Illusion – One and Many are the Same.
– Goethe

The endlessly creative possibility potential of EOn arises in *the Eternal Present Moment*. That which is perpetually birthing Itself from Itself, constantly in the process of *becoming* a new creating. This perpetually renewing infinite creative power makes all things anew continually.

It is like the beginning of a song that never stops singing. It does not sing in order to come to the end of the song, as it is expressing the Infinite to flow through the many and various media of finitude in the openness and flow of the present moment quantum field.

The Western esoteric tradition identifies this endlessly creative alchemical *EOn* as the *Ouroboros Serpent*, which continually swallows its own tail and continually sheds its own skin, symbolising endless connection contained within ItSelf. The Ouroboros is perpetually remaking and rebirthing ItSelf within It Self-impregnating and Self-wedding Itself.

This continual transforming and perpetual birthing feeds, fertilizes and engulfs Itself within a Self-contained cycle that transforms one energy into another, creating and simultaneously uniting the ideas of beginning and the end, alpha and omega, in an endless quantum spiral.

The Ouroboros is the *"set of things that are not yet things"*, the primordial origin. It is all that we know and all that we don't know. It is the precondition or blueprint for the existence of all different things, and where what can be done, what is possible, has not yet been put into form. It contains in embryonic form everything that can IN PRINCIPLE be possibly experienced, and the quality of consciousness that is the experiencing itself.

Ouroboros presents us with a set of paradoxical transformational abilities that can only be known by our embracing of the unknown. Everything we know is because we, or someone else, discovered it in the course of an encounter with something unexpected, and then shared it with others. This change brings potential for evolution and new opportunities.

These places of transformation and change, where something is happening that has not yet been modelled, where behaviours, laws and attitudes have not yet been erected, where something is happening that has not yet been understood, is where we can most thrive and evolve, yet is also what we can be scared of.

This is one of the gateways into *EOn*, which is why "left-handed" Tantric practices can be used to access this state directly and swiftly. Left-handed tantric practices directly confront your deepest fears and joys experientially and immediately: so, for example, a left-handed Tantrika would learn scuba diving if he was scared of deep water. Similarly, he would enter consciously into any and every situation/person/event he was scared of, in order to master the situation and transform the energy into bliss.

Polarities

Often we do not notice how opposing forces agree, and humanity often thinks he can sever that continuity and exist apart from it.[77]

In *EOn* the quantum blueprints of masculine and feminine principles form in all possible permutations of polarity, whole in their dance of union with other within the One Self.

EOn is where awareness witnesses its moving consciousness within the present moment. The fully Self-Contained Egg becomes impregnated by Itself by the still witness seeing its energy in infinitesimal motion. In so doing, the action repeats into infinite time from infinite time, ad infinitum.

These infinitesimal movements are the subtlest shifting of possibilities, forming infinite potentials within Its holographic container. Through these infinite reflections of itself, the mirror further repeats its reflections, all within the same quantum matter.[78]

This transformation is similar to how quarks move, the subatomic constituents of matter that appear and disappear in an instant, flashing in and out of existence like the manovrittis. Yet still, it is in the same consciousness, where the most ancient past and distant future are conceived of in all possible timelines within one awareness.

Every effect in Nature is divided into pairs of opposition. Each one of each pair is the reverse of the other, each a mirror reflecting the other. If one hand multiplies potential, the other simultaneously multiplies it equally.

– Walter Russell, *Atomic Suicide*, p. 288

The quantum polarities formed in *EOn* create the potential for *all* forms of coupling in virtually infinite combinations and permutations. These four sets of polarities that make eight in all are mentioned throughout the Sutra as:

1. Shiva-Sakti: First light of awareness and movement-desire
2. Maya-Isvara: Maya and Isvara
3. Brahma-Saraswati: Field and Wave of physical creation
4. Prakriti-Purusha: Matter and Light

In the human experience these four sets of fundamental polarities are present within every person as sexual orientations and connections, and are also reflected externally in all forms of relationship:

1. Male-female: positive negative
2. Male-male: positive positive
3. Female-female: negative negative
4. Female-male: negative positive

These four sets of twin polarities make eight, acting like an infinite battery, patterning, reflecting and feeding each other in the Cosmic Egg. They are the quantum blueprints of polarity perfected, each pair in constant infinitesimal action and response, in continuous flow without stasis or division, endlessly dynamic, always happening.

This is the blueprint and Seed of matter and the continuity of creation, a "fruit tree yielding fruit whose seed is inside itself."[79] These infinite batteries are perpetually birthing and revealing every possible potential in every possible permutation of every possible polarity in order to actualize and manifest the blueprint of every possible experience, process and form in the next Class of Sound *AIAUch*.

The Original Eight Cells

Man is formed in God's Image.

EOn manifests in our bodies as our first and original 8 Cells, revealing themselves after the fusion of Egg and Sperm. They form at the base of the embryo's spine within the first five days of conception.

These 8 Cells contain all the information (karma) for your incarnation patterns, the information from your previous actions and experiences, the reasons you have incarnated, your soul purpose, lessons and traumas, and all your remaining experiences, lessons, karmic patterns and causal wounds.

In each lifetime you have certain lessons, gifts, soul purposes, karmas and patterns that you play out and learn from. The unresolved ones you do not complete are carried on to your next physical body through your lightbody when your present physical body dies. In other words, the light body matrix of information held in the 8 Cells takes unresolved information upon the death of your physical body and migrates it to your next physical body in the wheel of reincarnation.

This wheel continues to spin, giving you lifetime after lifetime of experience and opportunities to resolve your karmas, conditionings and learnings until they are released from your 8 Cells. If these are not released, moved through, learnt from and resolved, you will continue to reincarnate through the information held in this 8 Cell Matrix.

The 8 Cells also hold the solutions to complete your incarnational lessons and awaken. Thus the 8 Cells hold your "destiny *pattern*" *for both your karma and your Awakening*. The 8 hold the information for reincarnation, which is set for each person *before* they incarnate as an embryo in the womb. The 8 Cells hold the reasons why you are here now on earth in a body, and are thus a huge Key to the dissolution of your karma and total Awakening.

When the 8 Cells are cleared of all patterning and conditioning, then we can live in the flow and awareness of the Blueprint Code of the One. There is then no need to continue reincarnating, as all limited patterns, informations and conditionings are dissolved.

The manifesting of the karmic seeds in each individual is found mathematically in the Phi Golden Mean progression, which started when the One made the One 1+1 *A+I*, and then the 2+1 in *AIUn*, the 2+3= 5 in *RLRK*, and now the 3+5= 8 in *EOn*: and thus is born the cuboctahedral shape.

This 8-celled cubic geometry is known as the *Egg of Life*, and stores the history of creation of the human being and the universe in this quantum pattern. These 8 Cells of *EOn* develop the creation pattern and sequence by forming these 8 original cells as a storage mechanism and holding geometry, where the One divides and develops Itself into these 8.

The informational patterns held in the 8 Original Cells are uniquely your own until they are cleared. This is why the maxim *"to thine own self be true"* is so relevant. Every individual has their own unique pattern for their life held in their original 8 Cells, and only by following this information and living it can you be who you truly are, and therefore free. Everyone's pathway is unique. The Science of the 8 Cells means the end of religion and the establishing of Sovereignty for each and every individual.

The 8 Cells are the constituent makers of us and the quantum matter universe, before matter in the 3rd dimension exists. They are our building blocks and form the quantum blueprint for life.

Egg of Life
First 8 Original Cells of Life

Egg of Life

3D Egg of Life
First 8 Original Cells of Life

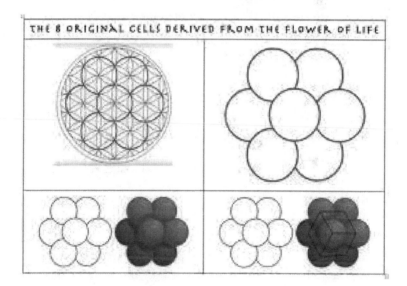

THE 8 ORIGINAL CELLS DERIVED FROM THE FLOWER OF LIFE

The Human Polarities

The processes within the cosmic egg are a direct reflection of our own birthing process growing as a foetus in the womb. In this verse of the Sutra, we have the word *Vijnanam* as being the intuitive insightful capacity for direct realization of reality. Vijnanam has many layers of meaning: it also means intelligence, consciousness, knowing and wisdom. On a foetal developmental level, it is the forming in the foetus of the faculties of perception (*saṃjñā*), feeling (*vedanā*), individual will (*saṃskāra*), consciousness (*vijñāna*) and form (*rūpa*).

Together, these faculties form our fundamental translating mechanism to perceive forms, and therefore all vibrations, as name and form, are one and the same polarity. Names are vibrations that are forms, and this *nāmarūpa*, the principle of names and forms as one process, is the guiding principle behind the architecture of creation. In *EON* this blueprint unfolds through the unified polarity of Maya and Isvara.[80]

We are formed by these processes, and like the One that moves throughout all life, we too are a mirror of Shiva-Shakti, which created us and the universe in order to experience and enjoy Itself. We too wish to enjoy all parts of our self, and we go about this by seeking to experience all aspects of our multifaceted masculine and feminine sides in many ways, through relationships, family, community, sexual preference, tantric practices and meditation practices.

It is only through direct experience that we can meet our polarities. We usually do this through external relationships that mirror our internal polarities, with partners, parents, children, family and friends. We look to relationships, to someone "else", to "something else" to mirror our internal polarities back to us, so we may see and recognise all parts of ourselves. This urge comes from *EOn*.

For example, we may sometimes be more masculine in one relating and more feminine in another. We are a son to a father, daughter to a mother, and vice versa. We can have a masculine body but be more feminine with another man or woman depending on the dynamics being played out, and we can have a feminine body but be more masculine, maybe seeking out a more feminine man.

We sometimes have a more feminine relating with another person, even if they are masculine or feminine, and sometimes both of you may be more masculine with each other. Sometimes having two polarities that are the same in a relating may get boring, confronting or conflictive, in which case we can switch our polarity within the relating, or leave the relating altogether.

We sometimes guide with one polarity and sometimes are guided by another, having relationships with others which help us recognise and experience our subconscious anima and animus, our hidden subconscious male and female urges and aspects. In so doing, we experience and learn more about our own masculine and feminine, and what needs attention within us.

We may try different modes of sexual expression: gay, lesbian, bisexual and heterosexual, *all of which are polarities within us* we may look to experience outside of us in order to internalise that pair of polarities. *We all have these polarities within us on the most fundamental level,* so to judge anyone else in their sexual expression is only judging an uncognised, unfulfilled and unmet polarity within your own self. Judging others' sexual preferences, expression and identity is a trigger for you to see which pair of *your polarities* are unmet, unrecognised and unfulfilled.

There are many combinations of polarities. We subconsciously look for and are attracted by ones that have a charge for us, a pull or aversion we feel to them. This helps to trigger aspects of ourselves into becoming aware of our own unmet, unresolved and unconscious polarities through these experiences.

Polarities form the primordial blueprint of all relationships. Fluidity and openness allow the cycles of masculine and feminine to support each other. This is the freedom to be totally masculine, the freedom to be totally feminine, the freedom to be both at different times, and the freedom to be neither of them.

This freedom to switch polarities and to be beyond polarities, to be androgynous, to receive and hold, to penetrate and direct, to be open and surrendered to the Primordial Root of Being, is to be able to express in all ways: freely, wildly, sweetly, lovingly, silently, receptively, penetrating and being submissive, guiding and surrendering. As we express all sides, we eventually enter immortal flow. We let go in all ways.

What is it to be Alive? is the question asked in *EOn*, with the answer: to be perpetually expressing Self, in totally free relating, within internal unity.

The journey of polarities is an emotional, primordial, preconscious, archetypal and soulful experience, as our emotions, relationships and sexuality are the main "charged" ways in which we most readily and immediately experience polarity. In Tantra, this journey of polarities culminates in the highest of Tantric practices: *Yab-Yum*, where "in the union of solar and lunar energies found in the two poles of the human energy system, negative and positive circuits are joined in a lightning circuit. A lamp can be lit."[81]

Yab-Yum is a practice that connects you with your partner on every level in order to realize the Self. Yab-Yum brings together all your mutual energies in polarity, breath and love, in every chakra within yourself AND with the other. In Yab-Yum each of you experiences, connects to, and integrates your own deepest polarities in order to cognise and realize the Self.

In Yab-Yum, two come together in One Wave. The yoni and lingam come together so totally that neither yoni nor lingam are felt separately anymore. Woman feels totally safe and trusting, giving of herself to open up completely, inviting her man in. All of the man moves into the entire woman and is totally received. Both woman and man feel totally seen in their essence and beauty, and this too is let go of.

In the authentic Yab-Yum joining, the two generate an energetic *Vesica Pisces*, entering the first movement of creation, the trinity *AIUn*. Here, the Trinity is of Shiva-Shakti-Human as One Awareness Energy Everywhere in One Body.

This opening into the Vesica Piscis then leads into a literal black hole, where both people dissolve who they were, into who they truly are, in a process of death of the separate self and resurrection of the infinite Self, as we shall discover later in the book.

Couples who have engaged in this sacred marriage, hieros gamos, sacred union and Yab-Yum journey, such as Christ and Magdalene, Buddha Padmasambhava and Yeshe Tsogyal, and many others, realize that form and emptiness, male and female, Shiva, Shakti and human, are One, having distinction yet having no separation.

One could say to the other, *"There was never a place where you were not. Your life from its very first appearance is part of the same flame as me, up to this very moment."*

Sounding EON

E is the first of the four sounds of *Kriya Shakti, the energy of action of the One A. E* is the first state of Kriya Shakti known as *asphuta*, or not vivid action. In *Eon*, we have A endowed with two out of the four Kriya Shaktis, with the four being E, AI, O, AU.

To sound E: Open the mouth and sound AH, raising the tongue and closing the mouth slowly until you hear E, just short of making the sound I.

O is the third Kriya Shakti, the third most vivid sound of action or *sphutatara*. O arises by combining A and U. The Clear Light of A, wed with all-pervading existence U, manifests the quantum blueprint for life in O, the seed sound for the egg-womb. In O, we are birthing and rebirthing in every moment in vivid action. O occurs when we surrender, when we let go and open into this deep presence.

With AH, O is one of the main sounds of sexual orgasm that naturally arises, regardless of race or nationality. It is a fundamentally primordial sound of vivid action. When we travel into and with the sound current of O and its associated sounds into our womb/hara and out through the crown chakra, we can access profound states of consciousness.

To sound O: O can be a short sound or a long, deep sound, arising from the labial U sound. Purse your lips together as if you were blowing out a candle and make the U sound: then open the lips more and round your mouth into an O shape. *U becomes O.*

N: N is Anusvara – the nasal N (as shared earlier in *A I Un*) that resonates the brain and pineal chakras, bringing inaudible vibrations into form. Adding N to O in *EOn* resonates this sound current and connects it into the pineal, helping to bring the formless into form.

EOn Sound Practices

1. Sound long *E into the pineal* for 8, long *O* for 8, long *NN* for 4.

2. Sound long *E* into the pineal for 8, long *O* for 8, long *NN* into the base of spine/perineum for 4.

3. Sound long *E* for 8 into the *base of the spine/perineum*, long *O* for 8, and *NN* for 4.

4. Inhale *E* into the throat; exhale *ON* deep into the sacral chakra.

   As you sound *O*, the sacral chakra contracts inwards. As you sound *N*, contract the perineum and sacral.

5. Inhale *E* into the pineal, exhale *ON* up into the crown.

6. Place your first and third fingertips to your thumbs. Sound *E* in your pineal. Direct the *O* sound from your pineal to travel back through the pineal channel into the base of your brain (just above where your spine and brain meet). Sound *NN* here. Now, sound *E* at the base of the brain, and use *O* to travel back through the pineal channel into the pineal chakra. Sound *NN* here. Repeat x13.

7. Sound *E* in your crown chakra, and then travel upwards and outwards with *E* into your soul star chakra, one hand's length above the crown. Do the same with *O*, and then make the whole sound *EON*.

8. Now, do the whole process internally. Hear each sound before you internally sound it; tune into its sound current that is already here. It may sound different, even angelic. Once you can hear the sound by extending your sonic imagination just a tiny bit, join in with it internally, sounding *with* this already existing sound current.

   So, hear *E* in your crown chakra. Travel up and out with this sound into the soul star chakra. Then join in with it, internally sounding *E*. Do the same with *O*, and then the whole sound *EON*. Surf it, and let go into this sound current.

AI AUch: The Song of Creation

ऐऔच् ब्रह्मस्वरूपस्सन् जगत्स्वान्तर्गतं ततः।
इच्छया वस्तिरं कर्तुमाविरासीत् महामुनिः॥ १४ ॥

ai auch brahmasvarūpassan jagatsvāntargataṁ tataḥ
icchayā vistāraṁ kartumāvirāsīt kṛpānidhiḥ

14

AI is the unified state of *A* and *I*,
formless awareness moving as Its consciousness.

AU is the unified state of *A* and *U*,
formless awareness shining in eternal time
as all-pervading sentience.

AIAUch is the true form of Brahma,
encompassing the whole moving universe within Itself.

Willing to bring out and expand
the entire range of the universe
from within ItSelf,
It willed to manifest Itself
in a shower of grace, love and blessings:
and the worlds were born.

Everything you see has its roots in the unseen world. The forms
may change but the essence remains the same.
– Rumi

In the first trillionths of a second when the physical universe
appeared out of the quantum field in the "Big Bang", the
quantum code became established as the underlying matrix of
the physical universe.

The fountaining forth of physical creation is the quantum
matrix emerging as Itself into Itself. This matrix is the release of
energy itself, i.e. the Big Bang did not occur somewhere in space –
it created space and occupied the whole of space simultaneously.
The One encompasses the whole universe within Itself.[82]

All the information is already here, there and everywhere,
established in the quantum code that is the only energy that is
existing everywhere already. Nothing moves as it is already all
there! It is instantaneous. Linearly, this is mind-boggling. In the
quantum world, it makes total non-dual sense.

Throughout the Sutra, creation unfolds through the unified
polarity of Shiva-Shakti in their various forms and functions.
As Shankaracharya says, *"when Adinatha (Shiva) and Layeshvari*
(Kali) catch a glimpse of each other, the world originates from them.
When their dance of Tandava and Lasya ceases, the world too ceases
to exist."

In *AI AUch*, the masculine and feminine see each other for the
first time *as forms*, enabling the birthing of the blueprint of all
forms in a shower of light, kindness, compassion and blessings.
It is from this *kripanidhi* that all creation arises from, activated
by the will of awareness and the desire of Shakti. Creation is
born from love, and this love is in all parts of creation as the
cause of creation.

AI-AUch is a mated pair that excite and activate the
potentialities of the quantum field into manifestation, or more

accurately reflect each other to bring forth the appearance of creation. *AI AUch* marks the coming into physical expression of this dance of polarities, each one of Shiva and Shakti retaining the same essence as before, yet each now taking on a different function, as marked by the change of name from Shiva to the creator function of *Brahma*.

AU is the sound of *Brahma*, the field containing and guiding the waves of creation,[83] the field which physical reality appears in, and the bridge between inner and outer experience.

AI is the expanding feminine creative wave, the seed syllable of Brahma's feminine complement *Saraswati*, Goddess of speech, sound, music and the Word that ignites creation. Saraswati here is the creative flows and waves of life, the manifesting sound current.

Brahma is the will to create, Saraswati the "singing" of this will into existence. Through their love, delight and desire, this field and energizing wave unfolds and manifests matter creation in a dynamic fountaining forth of forces, delivering form as these forces overflow in this Universal Song (universe means one spin or one song).

AI and *AU* are two of the most potent seed syllables in Sanskrit mantra. *AI* is the union of *A*, the clear ever-present light of awareness, with *I*, its desire moving power of consciousness: the first two sounds of the quantum code.

AU unifies *A*, the clear ever-present light of awareness, and *U*, eternal time and all-pervading sentience that sustains the quantum code in *AIUn*: the first and third sounds of the quantum code. These two conjoined vowel sounds activate quantum potentials that now fountain forth vibrations and matter to appear within Itself.

AI

Energy moves in Waves. Waves move in Patterns. Patterns move in Rhythms. A Human is just that: Energy, Waves, Patterns, Rhythms... a Dance.
– Gabrielle Roth

AI reveals the invisible quantum field through perpetually moving waves of vibration. AI or *AIm* is a feminine creative power and the seed syllable *bija* sound of Saraswati, the mother of music, speech, language and all mantras. As Vak she is the Word or Voice which sings the universe into manifestation from the quantum field in waves.

AI mirrors the multiple planes of the zero curvature of space-time geometry to create the appearance of motion through waves. These waves compose creation in every way, from music and speech to the dancer's grace of mudra and movement, and the art of blending smells. All of these, and more, are harmonic wave forms.

AI is the art, sound and science of communicating the infinite in waves. Saraswati is "she who continuously flows", and in this felt experience one becomes a transparent conduit for the magical currents and waves of creation. We become part of the whole wave.

As the consort of Brahma, she needs Brahma's field and directing will to "sing" the waves of all vibrations into forms in harmonic order and coherent harmony that flow in geometric whole number patterns.

Together with her partner AU, AI-AU dance and flow in infinite waves of desire fountaining forth joy, a delightful oceanic orgasm of blessings arising within a kind, still, singular awareness.

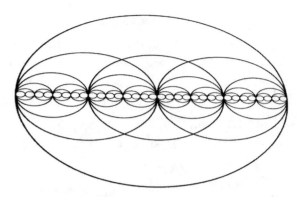

The wave is a spiral. A wave is not a single unit. It is a series of two opposed units. A wave has changing planes.[84]

Flow is a state of effortless consciousness – where things just happen through us. We access flow as our mind quietens, allowing our intuitive faculties to flow unimpeded. Having this experience shows us this potential, and once we are aware of this, it can be easily repeated so we can flow even more.

When we are in flow or in "the zone", we come into harmony with the waves of creation. In this experience, when speaking, creating, writing, expressing, you may have no mental idea what you are doing, or how the activity is happening. It just happens and expresses through you.

Being open to the wave brings forth energy and wisdom you have no idea you possess until the moment it comes out of your mouth or expresses itself in some way! This is an amazing experience to engage in, and one that many experience daily.

This natural flow can be accessed *at any moment*, by you being totally in the present moment. The greatest wisdom, synchronicities, feelings and connections occur when the wave flows through you.

There is less need to plan once you are in the flow of something greater than yourself. For example, if you are giving a speech or presentation, you may prepare something beforehand. Yet, if during the presentation you allow the wave to flow through you and connect with the other people in the group-wave consciousness, spontaneous and more relevant wisdom will come through you.

Your audience will join with you in a shared wave, where each one of you feels, receives and expresses more, as you are all part of waves in the same field. It is our natural state to be in this flow of the moment, in the natural flow of creating.

The Torus Spiral

AI is the creatrix of the torus movement and spiral that is *the one core recurring pattern in nature that develops and evolves life on every scale*. The torus has been observed everywhere, from atoms to cells, flowers, humans, planets and galaxies. It is the only energy pattern that sustains itself, and is made of the same substance as its surroundings.

The Torus is the shape, process and movement that exists at all scales in the universe, from the smallest to the largest. It defines the shape and movements of all life, as well as being the co-creator of the movements in the structure of space-time itself.[85]

It is the spiralling shape of all protons, the shape of the iris' lens of the eye, the shape that defines the movements of the human energy field, the shape that defines the movement of the north and south poles of Earth, the shape that defines the movements of our Milky Way Galaxy.

Each one of us is surrounded by, encased in, and penetrated by toroidal wave-fields which connect to every other toroidal wave-field in creation. Each one of us has this torus flowing in and out of us at all times, and the torus forms the development of the foetus in the womb, flowing into the nascent embryo, shaping it, and then flowing out, further shaping it, to then form our very first organ: the heart.

The heart is a "helicoidal myocardial band" that has spiralled in upon itself, creating its shape and separate chambers. This "Helical" Heart is a spiralling organ of perception formed by the torus,[86] and becomes the centre of the human toroidal fields that connect to the universal toroidal fields through 2 Golden Phi spirals that form an "eye" or centre point in the heart, which then emits electromagnetic fields outwards.

Heart moves blood, and our blood cells are flat tori. This toroidal spiral dance is in the bloodstream and in the blood cell itself. Blood flow is composed of two streams, spiralling around each other in a toroidal flow, looking like a DNA double helix, at the centre of which is a vacuum.[87]

A torus is double-ended and self-organizing, with the inside connecting to the outside; it is essentially an imploding sphere in constant movement. In its "doughnut" shape, wave-energy radiates fields of force, and when the limit of the field is reached, the energy spirals inward to the central source point. These streams of radially reciprocal spirals consist of two "tori" in masculine and feminine polarities, two opposed units with changing planes[88] complementing each other.

Each spiral expresses their qualities radially in all directions, constantly feeding back to each other in a radial dance within a circuit whose beginning and end are the same. This constant cycling into and out of itself keeps all vibrations and information continuously available to every part of the spiral.

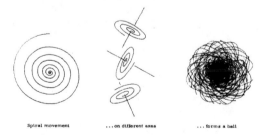

Spiral movement   ...on different axes   ...forms a ball

Streams of vibration rotate in spin networks, reflecting and clustering together multiple harmonics or seed syllables to form matter.

The torus is where matter and energy become an indistinguishable plasma, a singularity recycling continuously within itself – here, there and everywhere in its vortex at the same time. Torus or Shakti consciousness is a physicalised reflection of formless awareness, non-matter manifesting matter.

Professor Ashok Gangadean found the torus to be an accurate scientific model showing how each individual is completely interconnected to all of existence. This is the physical mirror of awareness: as above so below. All things arise from awareness: in the torus all things are interconnected.

Through the torus we can experience all space-time dimensions, as it is an immense self-contained spiral constantly generating intersecting energy patterns that weave, cross and re-cross to constantly re-create and sustain itself, developing the past, present, future and all dimensions. We can perceive the entire toroidal wave-field and all the dimensions it forms when we realize we are the awareness that it all appears within.

Recent discoveries have shown that the toroidal spiral is the fundamental mathematical matrix pattern of the structure of matter and the physical universe. This toroidal matrix has been discovered by author and scientist Lynnclaire Dennis, who has called it the "Mereon Matrix".

This matrix is six-tori inside five tori, which is "where non-matter manifests all matter." The knots of these tori are sound frequencies which have an accuracy of 6 decimal points, and her company has created various sonic programs to help people attune to these frequencies.

AU

AU is where the Will, Knowledge and Action of the One all combine. In Kashmiri Saivism, *AU* is where Shiva is most vividly existing *in manifestation*, in full action in the manifested universe.

AU in manifestation is the universal space-time geometric field that directs the wave patterns of the torus. Both are indivisible, with Saraswati *AI* reflecting Brahma *AU*, just as Sakti *I* reflects Shiva *A* as the reflected consciousness of awareness. In non-dual quantum physics, there is no movement: just reflections that appear to be moving and which appear to the senses-mind as waves.

AIAUch is the explicated and unfolded expression of the quantum nested state of *A I Un*. *AIAUch* is the same expression conjoined, transcribed and replicated into another order of complexity in order to bring about manifest creation from the quantum code of *AIUn*.

With the addition of the *anusvara MM, AU* becomes *AUM*. Anusvara is the dot above AUM signifying the *Nada Bindu* Singularity from where creation arises. The anusvara *MM* is a bridge sound that brings the formless into vibrating forms, bringing the breath of life forth from the formless.

When we add Anusvara to any *bija* seed sound, it helps lead the mind into one point and into silence. AUM is the active field in which all physical creation is held in and appears within, and the waves within this field are AIM.[89]

As you open your mouth fully, *AH* naturally arises. When you close your mouth whilst making sound *MM* naturally happens. With a half-open mouth in between these two, *U* is heard. *A* opens and *M* closes, with the midway point being *U*, the "all-pervading sound of existence".

A merges into *U* into *M* into silence. This is the same way sound moves within you, from the spoken word in *vaikhari* to the subtle internal picture sound *madhyama*, into the causal *pasyanti*, and beyond manifest vibration into the quantum field *Para*. Therefore it is said that the sounds of all languages are held in AUM, the sound-code of the activating field within which physical creation manifests.

The Four Faces of Brahma

In this verse the phrase *"brahmasvarupassan"* is used. This can mean two things: literally it means "Brahman, the totality of all that exists", which in the context of the Sutra is the quantum matrix emerging as Itself into Itself in *AIAUch*. The release of energy in Brahman created space and occupied the whole of space simultaneously as its own form. "The One encompasses the whole universe within Itself."

The second meaning is more obscure and refers to an ancient name of Brahma known in the *Rig Veda* as *Prajapati* or the first father of creation, a creator so ancient that he is now largely forgotten. This first father of manifest creation is the Seer or witness of all names/vibrations and objects. By Seeing all these objects and vibrations, the universe came into being.

In the creation story of how Brahma created the physical universe, as related in the Vedic and Upanishadic texts, Brahma arose out the navel of Visnu, seated on a lotus flower. He looked around him with his four heads in four directions into fathomless emptiness, and Seeing there was nothing he was unsure of what to do, as he knew his function was to create, but he did not know what to create!

Brahma then went inside himself, realizing he should express *what was within him*. Hence the term *brahmasvarupassan* is used, "the totality of all that exists", the quantum matrix emerging as Itself into Itself. This action created space and occupied the whole of space simultaneously as its own form, hence "the One encompasses the whole universe within Itself".

Brahma "saw" each and every name or vibration of existence within him, thereby naming or *nama* these forms *rupa* into manifestation.

In another version of the story, a lotus appeared from Brahma's navel and his consort Saraswati appeared. Entranced by her grace, beauty and elegance, filled with desire for her, his four heads lustily followed her as she danced around him. His desire for her was his desire to create, and between his *AU* and her *AI* waves of vibration issued forth into manifestation. (Indeed, the root sound *Brh* in Brahma means to expand, grow and breathe out, which now happens in *AI AUch*.)

As these waves radiated forth, they formed a *pragnanz* – a circle that is not actually drawn, but formed by the lines themselves. These lines are infinite emanations of pure awareness: the void in movement. In quantum physics these are known as Null lines, paths taken by light rays and massless particles. There is no difference between these lines and the void/awareness: the lines and circle are the same.

A geometry based on null lines is the Grail of quantum physics, for in a universe having such a geometry, mass does not exist. All distances, times and ideas of large or small, micro and macro, no longer exist, and no time elapses in travelling from one point to another. Not even one second passes from the time you leave, say Sirius, to Earth, for along such a null line the distance to the stars is 0. When you look along a null line, nothing separates you from all that you see in the universe. *In fact, you are everything that you perceive around you in all directions.*

This is how Brahma manifested names/vibrations as forms. He brought them out from within himself, and his four faces gave the field direction and geometric structure. The Brahma function created the template of space-time geometry by creating the architectural blueprints for the four forces of quantum gravity,[90] the strong nuclear force interaction, the weak force, and the electromagnetic force. These four forces work through the waves of the torus spiral, manifesting matter from non-matter.

AU is a resonant activating geometric structure known as the Vector Equilibrium or *VE* in physics: the only geometric form where all forces are equal and balanced, *"the invisible mother of all the shapes and symmetries we see in the world."*[91]

The Vector Equilibrium is the space-time geometry within which the toroidal waves of space-time curvature reflect. It is the geometry of physical reality in a Cuboctahedron shape which has 14 sides to it, the same number as the 14 classes of sound of the Maheshwara Sutra.

The Cuboctahedron has 24 edges and 24 triangular vectors inside it, and it is fractal because *the inside is the same as the outside*, as all 24 vectorial lengths are the same unit length as the outer 24 edges; no other shape in the universe has this property. This means that the external and internal perceptions are intertwined – inside and outside reflect each other perfectly at this point in creation, hence the saying "the inside creates the outside".

When the cuboctahedron shape is compressed it forms the 5 Platonic Solids that are the geometries of the five elements, the physical basis for all matter, as we shall see in the next Class of Sound that unfolds creation from *AIAUch*.

Brahma's four faces reveal the universe of forms or *rupa* through the power of vibration or *nama*, both of which need the energy of time in order to manifest forms. Brahma's four faces create the four yugas or great cycles of cosmic time, the macro version of Earth's four solstices and equinoxes we have four times every year. This also reflects in the four openings of time we have every 24 hours: dawn, twilight, 3:30am known as *brahma muhurta* or hour of Brahma, and midday.[92]

Brahma's four faces mirror his four arms: one left hand holds a cleansing vase, and the other the 4 Vedas. The right hand holds a rudraksha seed mala (the seed of Shiva) for mantra, with the other right hand holding the *brahma-tandram*, an oval disk with a beaded rim used to mark people's foreheads with their destiny.

This "destiny pattern" is recognised in *Jyotish* Vedic astrology as the information for reincarnation, which is fully manifested and activated within the foetus at this stage of their development in the womb *in AIAUch*.

Nama Rupa: Name and Form

In the beginning was the Word, the Word was with God, and the Word was God.
– *Rk Veda*, later used by St John in his Gospel

As the creation stories of the Vedas, *Bhagavad Gita*, ancient Egypt, the Bible and Kabbalah share, when "God" created the universe He named the sound of each form, thus bringing it into existence.

Adam named/activated each blueprint he "saw" to bring it into 3D manifestation. As Brahma looked within himself, he witnessed the blueprints or names *nama* of form and then vibrated each potential into form or *rupa*. These latent forms were manifested through the power of witnessing.

Nama or Name means vibration, which implies a boundary, a structure and a form. *Nama-rupa*, name and form, are one and the same, occurring together, not separate: one does not come after another for both are conjoined.

The true name of any form or *samjna* is that which holds all knowledge of that form and "to know all together as One". Samjna regulates the form, shape, function and the amount of energy *of any word, vibration or object.*[93]

The moment we name something, be it a person, quality or object, we give it a form. Name creates an image, a thought form, and we create images of everything by giving them names.

For example, by creating a fixed image of a loved one, which we all do over time and habitual experience, you restrict the possibility and potential of that person in your own mind, as you have already defined or named them by your image of them. You cannot be truly present with that person as they are already labelled in your mind as being like this or being like that. Your image of them has been created, and you relate to that image more than the person actually sitting in front of you. This can be hard to break out of as we see that person as a fixed image, rather than a living, moving process that can, will, and does change.

Naming things and people is a thought process. Thoughts come from words, which are names that define boundaries. Without words, there are no thoughts, no objects, no language and no way of communicating or naming anything in the third dimension.

Thought creates reality through naming something, which in turn creates a consensual reality, as we are all taught the same names for things. How could we ask another person to pass the apple without each of us knowing it was called an apple, and we all agreed this was so?

This is why the ancient Hawaiians, when the great sailing ships of Captain Cook appeared on their horizon, could not see them, as in their belief and language system, such ships could not exist. There was no name for such ships and thus they could not see them.

Thoughts are objects, known as thought forms. Without the names of things, our 3D reality would literally cease to exist as we know it. Without names, all forms dissolve into formless energy. This is easily verifiable by concentrated *dhyana* meditation on a form, or eye gazing. If you gaze long enough with focus into another's eyes, their face will literally dissolve in front of you!

The process of name and form begins for us when we are born into this world. We hardly have any cognitive functions and are largely unable to distinguish the different objects in our environment. As our minds develop, our cognitive function becomes stronger, primarily through words and speech. We are taught names for objects and so we are able to distinguish them.

We need names to order and discern our world. Our entire system of self-identity becomes built upon this structure that categorizes everything into separate forms, including our own identity.[94]

This accumulation of associations and experiences in names and words comes to constitute our body of knowledge and our sense of self.

Who we think we are is based on this infrastructure of names and forms formed on a subconscious level from our earliest years. Our name is our identity, how we relate, navigate and move in the world. Without having a name, it is hard to do anything in this world!

All names and forms initially arise from the showers of blessings and love of the One, as this verse of the Sutra states. Curiously, Brahma originally had five heads that became reduced to four, and legends around this point to some truths about name and form.

As it is said, as Brahma saw the gorgeous Saraswati sashay past him, he was besotted, desirous, mesmerized and enchanted by her beauty, grace and movements. He did not want to miss one moment of seeing her.

Brahma's fifth head formed at the top of his head to look skywards so he could follow and reflect the movements of Saraswati, the wave curvature of space-time. His tracking and reflecting of her wave planes allowed him to consistently hold an all-encompassing geometric field that contained and directed her fluid waves of vibration. This then bloomed forth the manifesting of the physical universe of names/vibrations and forms.

As the story continues, Brahma began to arrogantly believe that he was the only creator of the entire universe, rather than just being one part, one cog, one function of the One. His role was to witness the material universe of names and forms and thus bring it forth. Yet he forgot all that came before him and believed he did it all.

This was the first ego or *ahamkar* forming: the "I-Doer" which separates one thing from another. This was necessary in order to form the universe and our own sense of self-identity by differentiating one thing from another. Ironically, spiritual seekers seek to go beyond this stage of creation and their own I-self, yet none of us would be here or be able to Realize any of this without the ego ahamkar!

Brahma served the purpose of manifesting the separate self (the blueprint of which was formed by Maya before him) and differentiated universe of things and objects. Yet as a reminder to us of the necessity of cutting off the head of one's ego and its separating, objectifying, narcissistic self,[95] Shiva cut off Brahma's fifth head with his thumbnail.

This of course was the head that grew out to see the object of his desire Saraswati. Lust and the separate ego were created at the same moment as the universe was created. The result? Even to this day Brahma is seen as celibate or *brahmacharya*, and the Rishis and Kumaras he created were asked to reproduce *as he could not*.

There is another perspective to this story as shared in the Western Gnostic tradition, where the creator of matter and form is known as the *Demi-Urge*, a being who believes that he is the one and only God because he created matter. In his arrogance, he believes that matter is all that exists, forgetting about the formless awareness from which everything arises.

This arrogance is what Brahma is known for, and why Shiva, the formless awareness from which everything arises, cuts off his head and therefore his pride. The belief that form is all that exists limits humans to a mere 5% of reality, as modern scientists have now discovered with dark matter, which states that 95% of the universe is not form as we have believed form to be.

The belief that reality is *only* form has created much conflict, suffering, delusion and fear amongst humanity, and has separated humanity from their true nature. This is why Shiva cut off Brahma's head, to remind him and humanity of the formless nature from which everything arises, from which true freedom, happiness and Knowing arise.

Because of this, Brahma became relegated to obscure status throughout the later Vedic texts, culminating in him being the forgotten part of the Brahma-Visnu-Shiva trinity in modern day India, with only five Brahma temples consecrated to his name, as compared to the millions of temples for Shiva, Sakti and Visnu.

Brahma's main function was Naming and Forming the quantum potentials in *AIUn, RLRK, EOn*, and curiously enough this is the very function that Brahma has been obscured by, as his name and form have been relegated to obscurity!

This paradox of Brahma is understood through his very function. Name and form are the most subtle, pervasive functions in our minds and world. It is *the* integral part of how we think, act, speak and experience. Name and form is the foundation of our world, such an integral part of our infrastructure that we take it for granted and forget about its importance.

Without name and form, we could not speak, think, pick things up or do anything. Indeed, without name and form we would be amorphous blobs in a sea of quantum energy. We are so used to the names and forms of our world as the basis for our own self-identity and ego that we take it for granted.

As the function that names and forms everything in existence, Brahma is intrinsic within us, present every time we speak and think. "He" is always and already here *as this function* within us from the time of *AIAUch*, in both the cosmic process of forming the universe, and at this stage of our embryonic development when these faculties are formed within our nascent foetal structure.

Sounding AI AUch

The four faces of Brahma witness, explicate, unfold and direct the quantum code of the One, blooming forth the blueprints of all names and forms through the universal sound current *AIAUch*. Toroidal waves of consciousness flow throughout creation, throughout all sentient beings, interconnecting all life forms. This sound current is always and already here everywhere in creation.

Bija seed syllables are the building blocks of speech and sound, from where all speech and sounds arise, and when seed syllables are used correctly they can help you access universal sound currents.

For example, when one sounds and delivers syllables into the centre of a chakra, the chakra entrains with this essential sound frequency. In the process, shadows of the chakra will be felt and will have to be worked with.

The end result of this work is that when each chakra is touched, it resonates in joy, laughter and love, the "shower of blessings and love" mentioned in this verse of the Sutra. This "shower of blessings and love" is where all seed syllables arise from.

AI is an expanded vowel sound, long and weighty in pronunciation. Open the mouth wide as in *AH*, tongue relaxed, and add *I*.

Making this sound helps to ignite and focus creative energies. Adding the anusvara MMM we get *AIM*. *AIM*, like *AU*, is a Kriya Shakti sound, an energy of action. It is a sound that gets things done.

It is a motivating sound, and can be used to direct and strengthen creative willpower. With clear intention, it can orient you towards manifesting specific desires, and can evoke wisdom when you sound it with intention to know a specific subject. It can develop the power of concentration and can awaken intelligence in the *ajna* pineal chakra.

AU is an expanded vowel sound, long and weighty in pronunciation. Open the mouth wide as in *AH*. *AU* is the first vowel to arise as the mouth and lips contract and purse in order to sound *U*. When you hit this sweet spot, keep sounding it long, expanded and with some weight behind it.

AI AU Sound Practices
Start with sounding these mantras aloud with breath, and then progress to sound them internally.

1. AI AU
Inhale *AI* into the ajna pineal chakra.
Make long, slow tones that vibrate nasally.
Exhale a long *AU* into the ajna pineal deep in the brain.
Let it reverberate here.
You can use *MM* after *AU* and *AI* to resonate your pineal. Try with *MM* and without. You can also use the full sound *AUch*.
Repeat this x9 aloud, x36 silent.

2. AI AU
Inhale *AI* through the vishuddha chakra – throat.
Make long, slow tones and vibrate the throat.
Exhale *AU* into the ajna pineal deep in the brain.
Let it reverberate here.
You can use *MM* after *AU* and *AI*. Make *AU* long and use *MM* to resonate the space in the pineal or cave of your brain. Try with *MM* and without.
Repeat this x9 aloud, x36 silent.

3. AIM AUM ONG
The Tantric *bijas* of *AI AU* are *AIM and AUM*, sounds igniting the manifest universe. They are the simplest and most powerful

vowel sounds common to all languages, and move throughout all of creation. *AIM* is the Wave, *AUM* the Activating Field, *ONG* a creative moving power like *AIM*.

Inhale *AIm* through the vishuddha chakra – throat. Exhale a long *Aum* into the ajna pineal deep in the brain. *Let it reverberate here.*

Sound *ONG* through the medulla at the back of the neck where the top of the spine and brainstem meet.

Repeat this x9 aloud, x36 silent.

4. Aham Brahma Asmi: I Am Brahma

This is the moola-mantra or root mantra of Brahma. Below are three ways you can sound this mantra. Start with sounding them aloud with breath, and then progress to sound them internally.

1. Inhale *AH* into the ajna pineal chakra. Exhale *HA through* the vishuddha throat chakra. Continue with *"MMM"*, in the ajna pineal chakra.

   Do not vibrate the lips. Direct MM into the pineal deep in the brain.

   Inhale *BRAHM-AH* through the vishuddha throat chakra. Inhale *ASMI* into the ajna pineal, resonating *MI* nasally and into the pineal deep in the brain.

   Repeat this cycle x12 or more. Then sound internally.

2. Inhale *AH* into the anahata heart chakra. Exhale *HA*, focused on the vishuddha throat chakra. Continue with *"MMM"*, in the ajna pineal chakra.

   Inhale *BRAHM-AH* through the ajna pineal chakra. Exhale *ASMI* into the ajna pineal, resonating *MI* nasally and into the pineal deep in the brain.

   Repeat x12 or more. Then sound internally.

3. Inhale *AH* into the ajna pineal chakra. Exhale *HA* through the ajna pineal chakra. Continue with *"MMM"*, in the ajna pineal chakra.

Inhale *BRAHM-AH* through the ajna pineal chakra. Exhale *ASMI* into the ajna pineal, resonating *MI* nasally and into the pineal deep in the brain.

Repeat x12 or more. Then sound internally.

हयवरट्

HA YA WA RAt:
Space, Air, Water, Fire

भूतपञ्चकमेतस्मात् हयवरण् महेश्वरात्।
व्योमवाय्वम्बुवह्न्याख्यभूतान्यासीत्स एव हि॥ १५ ॥

bhūtapancakamētatasmāt hayavaranmahēśvarāt
vyōmavāyvambuvahnyākhya bhutanyasīt sa ēva hi
15

Mahēśvara wills into manifestation
the five elements *ha ya va ran*, space, air, water, fire.
He as Awareness is in these elements.

हकाराद् व्योमसंज्ञश्च यकाराद् वायुरुच्यते।
रकाराद्वह्नि[रेफाद्वह्निश्चक]स्तोयं तु वकारादिति शैवराट्॥१६ ॥

hakārād vyōma saṁjnanca
yakārādvāyuruchyatē
rakārādvahnistōyaṁ
tu vakārāditi śaiva vak
16

Ha is supreme space, the highest speech,
where the name and shape of all forms,
vibrations and objects are known as One.

Perceiving, seeing and discerning these
are the qualities of Air, *Ya*.

Ra Fire travels and arises through Water, *Va*.
This is the original *saiva* word and lore.

HA: Space

Space is the body of Self. Truth is its being, wherein the vital force disports itself, mind delights and peace and immortality abound.
– Taittiriya Upanishad 1:6

From the indivisible primordial quantum field of *AIUn, RLRK, EOn* matter creation erupts through toroidal waves of consciousness in the space-time geometry of *AIAUch*. Now comes forth the elemental superstrings of creation as an expression of these waves of consciousness formed in the space-time geometry field.

In quantum physics, superstrings are vibrating elastic loops of different mass that stream across the universe, creating, weaving and interconnecting all of creation through their vibrating strings. With our senses and modern scientific instruments, we define an aspect of these superstrings as the five fundamental particles of nature. These five types of particle emerging from *AIAUch* are related to each of the five elemental superstrings of sound.

Each vibrating elemental superstring of sound births the next elemental superstring in the unfolding sequence of creation. For example, adding another set of frequencies or chords to the first element created, Space, then produces a new harmonic set of sounds that becomes the next element, Air. This new cluster of harmonics then unfolds the next fundamental particle type and element of Fire, and so on.

The quantum spin types are:[96]
Spin 2 Graviton: Vyoman-HA-Space
Spin 3/2 Gravitino: Vayu-YA-Air
Spin 1 Forces: Tejas-RAt-Fire
Spin 1/2 Matter: Apas-WA-Water
Spin 0 Higgs: Prthivi-LAn-Earth

Space is the first element of matter to arise from the great wave of creating *AIAUch*. At this magical transition, there is a jump from the invisible quantum field to the explication of this field that can be measured with the senses-mind, as found in the Planck Length.

Space is connected to the element hydrogen, the first matter created from the Big Bang. As huge clouds of hot hydrogen gas emanated and then cooled, they formed the beginnings of physical matter. In the Periodic Table hydrogen has the number 1.

The word used for space throughout the Vedas and in the Maheshwara Sutra is *vyoma*, meaning ether, sky and the heavens. *Vyoma* has many other meanings, one of which comes from a root meaning "to cover or clothe". Vyoma clothes the infinite and formless with the first subtle substance, the subtlest form of matter, in order to express and manifest Itself, and these clothes are made from this substance.[97]

Vyoma holds all sound vibration in its cloak, as all sounds in creation *resonate in* space.[98] *Vyoman* is found throughout the Vedic texts, and remembering that the Vedas give keys to fully comprehend the Maheshwara Sutra, we find in *Rk Veda 1:164, v. 34*:

"Prcchami vacah paramamvyoma."
"I looked into the supreme space and highest heaven of speech."

"Brahmayam vacah paramam vyoma."
"The Creator Brahma holds the supreme space and highest heaven of speech."

Indeed, the mantras of the *Rk Veda* come from, and are held in, this highest space of Vyoma, where the gods have their dwelling place.

"Rco aksare parame vyoman, yasmin deva adhi visve niseduh."

In verse 41 the divine speech or the Word is described as "*sahasraksara parame vyoman*" meaning that the thousand syllables of the supreme speech or Word arise from the highest heaven of vyoman.

Vyoman is the supreme space, the highest frequency of speech where the gods dwell. These gods are clothed in mantra, made of mantra, and are revealed through mantra. In *Vyoma*, we can access the space that the gods are made of, in the substance the Word ripples creation into being through.

Parama Vyoman is "the ultimate and infinite source"[99] of space and is part of the Supreme Self. This is echoed in the Taittiriya Upanishad, where "the Knower of Brahman attains the Supreme in vyoma."

> "*Yo veda nihitam guhayam parame vyoman so'snute sarvan kaman saha.*"
> "Who Knows this, which is deep in the womb of supreme space, attains all desires."

All physical forms arise through waves in space. Matter is a spatially extended wave flow. Waves moving through space are vibrating, meaning matter arises from wave patterns flowing through space. These wave motions through space create matter through the vibratory mediums of the elements.

Space is the subtlest form of matter, the one substance that is everywhere in physical creation, connecting every point in the universe to every other point. For example, if you travel long enough in any direction you will eventually return to the point you originated from because of curved space-time, which eventually doubles back on itself. Space holds time within it, and the space-time continuum mirror connects all times through the wave structure of space, a process of consciousness and structure in union.

Space is the wave medium from where every material thing emerges through space's principal quality: sound. This manifestation is not something that happened once a long time ago and was completed. In space, subtle matter is emerging *in every moment*.

Space is the connective medium for all creation, a substance within you, around you, and within every object. There is nowhere it is not in creation. Space is the subtlest substance, and it is from space that the many things we experience arise. The senses cannot perceive it, taste it or see it: it is only through sound leading into silence that it can be experienced directly.

Space is full, not empty: full of light, sound and vibration in a subtle form, beyond the reach of the senses. Space is the subtlest form of physicality, filled with vibrational intelligences that manifest matter through air, water, fire and earth. Our brain interprets these vibrations through the senses, with all sounds, tastes, touch, smells and sights being vibrations held in potential in the field of space.

Most of who "we" are is space, found at the heart of the atom, around which swirl the ingredients for physicality. Physicists describe this as "empty space" only because they have not consciously experienced or meditated in space element.

Meditating in space element is like immersing into a clear, shimmering light full of the vibratory potential for all colours, sounds, elements, senses and matter of all kinds. It is a high, peaceful frequency everywhere in creation, and is the most transcendent element to immerse into, as it is the source of all other elements.

Everything in creation is held in the embrace of space, for space is the wave medium in which matter manifests. In this sense space is an interpreter, unfolding subtle harmonics from one wave pattern to another, bridging the sensual world to the mental and causal world,[100] a translator from the subatomic realm to the atomic.

Quantum physicists have now confirmed that sound, the principal quality of space element, is an interface and translator between quantum states, i.e. sound waves let quantum systems "talk" to one another. The University of Chicago and Argonne National Laboratory share that different types of quantum technology communicate to each other using sound,[101] connecting quantum states of matter.

To understand this property of sound, they use "spins" – the property of an electron that can be up, down or both. Transferring this information requires the translator of sound waves, which when coupled with the spins of electrons make these spins "pay attention".

In these quantum systems, every particle in the system is harnessed, which then grows exponentially. Sound connects and brings them all together, allowing the transfer of information to/through every part of the system, *unifying every particle of the system into flow.*

Space and Sound

Geometry is the study of structure in space, and all sounds resonate in and come from space. Space gives rise to sound and all manifest vibrations, which gives rise to hearing and speech. Space is the primary element of our *vishuddha* throat chakra, whose symbol is a sixteen-petalled mandala of the sixteen *matrka*, the sixteen vowel sounds of light that measure out all of creation.

The throat chakra is the medium, translator and expresser between the inner understandings we have of life in *madhyama* and the outer expression of this understanding through speech or *vaikhari*. The throat is also the channel for the ascending stream of *Udana Prana* which is a gateway to space element and the stream of prana the spirit uses when leaving the body upon physical death.

161

Curiously, we usually have no sensation of the presence of space, despite its existence all around and within us, because we usually only sense *energy transfers, not space itself.* Yet as Patanjali observed, we can tune into pure space element through deeply listening to and tracing a sound back into its source: space. When what we hear is united with its source, one enters this field.

When one focuses on this intrinsic relationship between sound and hearing, one acquires supra-sensory hearing or clairaudience, and enters space. The experience of space opens, silences, expands and makes you feel weightless, and has a specific field of light.

Sounding HA

Space cannot be purified, as the other elements within us can be: it can only be increased or decreased. We can become more spacious, or less spacious. Whilst space is about 6% of our physical constituency, the more we consciously develop space element, the higher this percentage increases. We can increase the amount of space element within us most effectively through sound and meditation.

The more spacious we become, the more our perceptual range expands *with* an increasing mastery of timeline navigation, as time is held within space. One feels vaster; the mind becomes fluid and quieter. The mind opens, affecting the body, relaxing and putting it at ease. It can be deeply joyful to be in space element, full of light. We become more accepting, more able to access the witness that can know all things and all perceptions.

The sound for Space Element is *HA*, the first of the semi vowels, linking the vowels of light (used in the previous four classes of sound) with consonants. Consonants create forms, and space is the bridge between light vowels and form making consonants, between formless and form.

HA is a sibilant sound, coming from the open mouth sound *AH*, injecting more breath through the *H* sound, with focus and attention on the throat chakra as the place where it emanates from. You can elongate the *AA* to create *HAAA*. You can also add the anusvara *MM* to *HA* to resonate the throat chakra more, making HAAAMMM, with the MMM consciously vibrating against and in the throat chakra.

Inhale *HAAAAMM* into the throat chakra for 9,
and exhale *HAAAMM* out from the throat chakra for 9.
Repeat 9 times.

AHAM MAHA AM

In the *Paratrisika Vivarana* of the famed Kashmiri Saivite sage Abhinavagupta, he describes how Self expands into creation through space Vyoman in the sound AHAM: I AM. As *AH* expands, *HA* forms. When this is reversed the word *MAHA* (great) forms as its mirror, making the mantra *AHAM MAHA AM*.

MAHA mirrors AHAM back to itself. This mirror reflects the very structure of the space-time continuum and its wave-planes. As physicist Walter Russell shares, "the planes of zero curvature which bound all wave-fields act as mirrors to reflect light from one field into another," yet Abhinavagupta knew this 1000 years ago! He Realized a mantra that can reflect this experientially for us, with this mirror of mantra and breath uniting the heart and mind.

Inhale *AH* for 7 into the ajna pineal.
Exhale *HA* for 7 whilst focused on the ajna pineal chakra.
Make the sound "*MMM*" focused on the ajna pineal.
Inhale *MA* for 7 into the anahata heart chakra.
Exhale *HA* whilst focused on the soul star chakra above crown.
Make the sound "*MMM*" for 7 focused on the ajna pineal.
Repeat this cycle 7 times.

YA: Air

The transformation into this form or that form is not driven by the causes near to it, just orientated by them, the way a farmer diverts a stream for irrigation.
– Patanjali's Yoga Sutras, 4:2.3

Space is excited through sound vibration into air, wind, *vayu or prana*: energy and movement that include the element of oxygen but is not limited to just the element of oxygen. The manifesting of prana from space activates, moves and breathes the living potential of space into manifestation. This moving of the winds, this exhaling from space, brings physical life.[102]

This breathing orientates the living substance and field of space, creating currents, flows, eddies, vortexes, channels and streams to express space's manifold potentials. These currents irrigate and nurture creation by forming pranic fields of moving, shimmering energy within space, acting as an attractor force, like gravity.

Pranic fields harness, activate, organize and move the creative potentials of space into manifestation. *Vayu* begins by unfolding spiral galaxies in a wave cascade of spiralling motion circulating harmonic vibration.

Vayu carries vibration, governing the moving of all things through expansion and contraction, in processes like the beating of the heart, the pulse of life and the processes of breathing. Vayu moves the elements, fluids and every state of matter around continuously as the moving life flow in all beings.

Everything moves because of air, which constantly circulates, energizes, connects and communicates different parts of our bodies, minds and spirits together. Vayu completes its manifestation within us in our five vital winds or five pranic flows, which activate, empower, release, raise up, descend, digest and connect these flows within us. These five pranic flows fuel and activate parts of our brain, known as the Trinity of Mind: *manas, buddhi,* and *ahamkar.*[103]

Vayu is creative and connects seemingly unrelated things. Air connects to the heart chakra as the sound YA, the sense of touch and being touched, being felt and feeling others. Air is also connected to our neural sense of handling and understanding, grasping and letting go, and is how mind and memory operate.

As Prana, it follows your awareness: whatever you focus on and place your attention onto, there Prana flows. Too much air or unbalanced air element within you can lead to one being scattered, airy or blown away like a leaf in the wind, taken here and there.[104]

Rishi Vasistha shares more about the direct experience of Vayu consciousness in the *Yoga Vasistha 6, 2:*

> By means of *Vayu Dharana,* concentration on the air element, I became the element of air. I taught the grasses, leaves, creepers and straw how to dance. In pleasure gardens I carried sweet odours, in hell I carried sparks of fire.
>
> My movement was so quick that people said mind and wind were brothers. I flowed with the waters of the Ganges, I assisted space by carrying waves of sound. I dwelt in the vital organs of all living beings. I knew the secrets of fire and became known as the friend of fire. I operated the bodily functions of all embodied beings and was therefore their friend and enemy at the same time.
>
> As Vayu I performed six functions: gathering together, drying, supporting, vibrating and causing all motion, carrying smell, and cooling things down.

Sounding YA

YA is connected to the heart, and its element Vayu-Air. Focus on the heart as you sing it. To sound YA, open the mouth wide with Y and add air through A. To make this more powerful, add *MMM* to *YA*, and elongate the *YA* sound. Make the sound from your heart: *YAAAMM*.

Inhale *YAAAMM* into the heart for 9,

and exhale *YAAMM* out from the heart for 9.

Repeat 9 times.

VA: Water

Hydrogen and Oxygen combine to form water: H_2O. In one theory of how water helped form life on earth, around 3 billion years ago, the first primitive life forms on earth evolved around hydrothermal ocean floor vents where water came into contact with lava and mineral-rich gases.[105]

Above these waters wind currents rippled the oceans' surface into turbulent waves. These rich, bubbling sounds imprinted structures in water's membrane surfaces called nodes, and in areas of high intensity called antinodes.

Fluid forms began to develop from these patterns in the harmonic constant of Phi, the Golden Mean of 1.618 found in all living things that organizes and structures life's development patterns. These sound waves created the initial patterns, or sonic infrastructure scaffolding, for more solid forms to manifest, as the science of Cymatics shows us.

VA or *WA* is the syllable for the water element. Our body is over 75% water and most biological reactions occur in water. Sound travels five times more efficiently through water than through air.[106] Water is informed and shaped by sound vibration, and sound is how every living being, from fish to reptile to human, came into existence on our planet.

Water is the perfect interface for vibrational information, as it moves through waves (like space as it reveals forms) and is present in every form of life: we all need water to exist. Information is easily transferred via water's crystalline structure, as water is a liquid crystal almost always found in a geometrically clustered state in the body.

Crystalline structures act as transmitters of information through the brain, DNA,[107] nervous system and cerebrospinal fluid, where the most structured water in the body lies, micro-clustered cubically, pentagonally and hexagonally.[108]

Scientists Hameroff and Penrose say structured water "could take on aspects of a quantum computer in preconscious and subconscious modes," meaning water acts as a memory carrier and transferring medium. Water records and transmits history, a liquid tape recorder for our feelings and thoughts. Water has memory of our origins, memory of our emotions and memory of our essence.

AMBU

Air brings forth water, and water arises from air: *YA VA*. Water becomes vapour[109] to feed our oceanic ecosystems and weather in the mantle just above earth. Air and water mingle in this intermediate atmosphere, moving according to the laws of liquid flow and fluid dynamics.

In air we find many movements similar to water, though at greater speeds. Both have eddies, flows, channels and vortexes. Water repeats its patterns, whereas air changes its direction and flow. Together, their interplay gives rise to new patterns and creates *a fertile atmosphere* highly conducive to the moulding and shaping of forms.

As air changes in density it becomes lighter or heavier, rising and falling, creating a rhythmic breathing through the atmospheric mantle as a vehicle for sound vibration. Water pressure contracts any form, and contracts the gases within any form. Its pressure is the same in all directions, and at supercritical conditions like the primordial processes of creation, water is completely miscible with gases like oxygen.

In the *Tattvārthasūtra 3:1,7*, atmospheres support each of the seven worlds. Seven atmospheres exist, one below the other, with *Ratnaprabhā* supported by the humid atmosphere *ghana*, in turn supported by a dimension of densified air and water *ambu*, which rests in a ring of thin, rarefied air *vāta*, which in turn rests in space: vyoma or *ākāśa*.

In the Maheshwara Sutra the word *ambu* is used for water, meaning water flow, coming from a cloud, moving in the water, water borne and the god of the waters. Ambu is a dimension of densified air and water, a dimensional strata or container they both converge in that is formed by their complementary movements, where they dance, intermingle and create patterns, fluid shapes and organic moulds.

Ambu is a shape making, form moulding intelligence which coalesces and coagulates currents of water and air together within a specific atmosphere (remember that Vyoma or space is an atmosphere as well) and then forms a strata or layer within this atmosphere where fluid forms are moulded and take shape.

The true name of a thing is known when we see it as a moving form.

Famed Austrian scientist Viktor Schauberger shares that *"water teaches us that what we think is solid is just a mass of different movements." Water is rhythmically intelligent, and the more water is rhythmically active, the more it is alive.*[110] Every action on the surface of water evokes a rhythmic series of waves, and between any two streams there is a balancing dance of vortexes.[111]

At the surface of contact between two movements, one layer rolls in on the other. Fluidity perceives differentiations, evening them out in a rhythmic process, forming fluid shapes in this dance. These moving formative processes are similar to those that shape our bodies in the fluid element, from the ear, to the brain, and the sense organs. In these fluid dynamics, the fundamental principles of organic development occur when many movements flow into, over, and through one another at the same place in space-time. In other words, *forms arise out of air and water through movement.*

As sound waves in fluid move forward, they collide with molecules and push them closer together. As the waves move backward, the fluid has room to expand and its density decreases. These movements mould fluid shapes, which then dissolve, releasing the form created.

These forces then appear as a function within the form just created. Thus, "solid" forms arise from fluid streaming forces that continually take shape and pass away, rising and falling, appearing then disappearing.

Rishi Vasistha experienced the multidimensional nature of water element in the meditation of *jalan dharana*, the one pointed concentration on the water element, where he:

> *dwelt in the bowels of the ocean, making the appropriate sounds. I dwelt in the bodies of plants and creepers, making my own channels with them.*
>
> *I entered the mouths of living beings and mingled with their vital organs in their bodies. I flowed restlessly in riverbeds and took my rest in the dams. Rising as vapour, I entered into the heavens as a cloud. There I rested with my friend the lightning.*
>
> *I dwelt in all beings as the water element even as the infinite consciousness dwells in all beings. Coming into contact with the taste buds in the tongue, I experienced different tastes – surely that experience is pure knowledge.*
>
> *When the flowers blossomed, I descended upon them as dew, and tasted whatever sweetness was left in them after the bees had their share. I dwelt in all the classes of beings as awareness of taste. Assuming the form of droplets, I enjoyed riding on the wind and travelling from one place to the other.*

The Water Element In You

Water is "a kind of love that holds everything together in creation in unity".

Water acts as a medium between opposites, a mediator between contrasts, between the solid earth and gases of the sky. Water flows through polarities, between solidification and evaporation, always retaining possibilities for transformation in any moment, constantly creating something new, dissolving what is rigid, bringing it to flow.

Water is nature's recycling system, flowing from liquids and solids to gases. It is matter in an ever-changing state. Water is the mixer of life. Its currents channel vortexes of energy, carrying information, receiving and circulating radiant energy, recharging and sustaining life.

Water nurtures, rejuvenates and harmonizes our emotions or *Rasa*. Our emotions need to be felt, expressed, released, flow and circulate. Through devotion, emotional healing and tears, we flow more with the water element and come into resonance with this natural order of creation through our emotions and feelings.

Swimming, deeply relaxing in a salt bath and walking by the ocean is an enjoyable past time for many, as it reminds us of our origins. This primordial memory of flow allows us to relax and expand into fresh possibilities and potentials.

Water is flow, a term developed by Dr Csikszentmihalyi of the University of Chicago, meaning the process of completing naturally and easily an action that we usually find to be difficult. As we become more fluid, the linear or static mind recedes into the background and we allow our intuitive faculties to flow unimpeded.

Flow is being in a state of effortless consciousness – where instead of trying to force something, we just allow it to happen. Having this experience shows us what our potential is, and once we are aware of this potential, it can be more easily repeated and flow more. Flow in our bodies expresses through these spiralling currents that effortlessly translate one form of information to another.

To be free, unrestricted, and to express yourself fully is being in flow. When we are fluid, we feel and can let go of everything and anything as it arises – we become transparent. Static, rote facts and figures knowledge is not the way of water, which is a fluid intelligence, not rigid or dogmatic. It is a "heads up" process, seeing what is in front of you and adapting accordingly, in the moment.

Fluid is a resonant element, meaning that fluid in the galaxy, fluid in the planet, and fluid in our bodies is engaged in a resonant stream of bio-cosmic nourishment.[112]

Our bodies arise from fluid streaming forces, which continually take shape and pass away, rising and falling back into the ocean of vibration.[113] That which is in a state of rest originates in movement, *and what from far away looks to be a structure, upon closer observation, is actually a process.* Process and structure unite in the moment, to then dissolve to form another moment... and so on.

Creating is movement. Your own movement or vibration is an oscillatory motion, and your frequencies occur in sound waves. In fluid, sound waves travel as a compression pressure wave, meaning that the speed of sound in your liquids moves through the force of hydrostatic pressure.

Your sound or acoustic pressure deviates from atmospheric pressure, which is caused by a sound wave, propagating further sound waves to travel through the liquids in your brain, heart and lungs. Blood pressure, moving through your veins, are sound waves or compression waves that create disturbances, contorting in the direction of this force, affecting your cells.

Pressure changes resonance. Sound waves cause disturbances in molecules, spreading from one to the other in a wave pattern. This sound wave pattern carries vibration from place to place within you, and is the repeating "pattern" of your pressure, compressions and refraction in your world.[114]

When the tissues in your body become rigid and dense they interfere with this natural flow of information from cell to cell, between your body and the larger fields we are a part of, resulting in us losing our fluidity.[115] As we slow down, relax and melt old patterning and habits, we resonate with this vast field of fluid intelligence.

Sounding VA

WA or VA is a labial sound. Sound the vowel U, with your lips pursed as if blowing out a candle. Now, open your mouth and lips into AH, and you will hear the WA sound. Pay attention to your lips in this transition and the WA will be clear. You can use either VA or WA sounds. Feel which one resonates with you most.

Inhale VAAAAMM into the sacral chakra below the belly button at the pelvis, for 9, with attention on the sacral.

Now, exhale VAAAMM out from the sacral chakra for 9.

Repeat 7 times.

RAt: Fire

Water is the source of fire. He who knows this becomes established in himSelf.

– Mantra Pushpam: Flower of the Vedas Chant, Taittiriya Aranyakam, Yajur Veda

Fire is a process, and the word used in the Maheshwara Sutra for fire is *Vahni*, used to highlight the intimate alchemical relationship between water and fire. In the *Rk Veda*, Vahni is the one who conveys, rides or is borne along through fire, such as a charioteer or rider, a conveyer and bearer of messages and offerings to the gods as the *flowing, streaming one*. (*Rk Veda*, ix 9, 6) Fire is flow, carried along by the flowing currents and streams of air and water, *Ya Wa*, in one fluid process arising from space.

As we have seen, fire at the bottom of the oceans provided a spark for creation. Fire breathes within water, igniting the primordial waters of creation into manifestation. Fire *RAt* emerges from water *VA*, and fire moves within water as *Pitta*, one of the three fundamental body types in Ayurveda.

Fire transforms water into steam, and fire with superhot gases creates fluid plasma as part of the creating process. Fire is also an integral part of the pressurized atmosphere water and air creates in *ambu* that brings forth fluid forms.

The subtle measure or explication *tanmatra* of the fire element is *rupa* or form, and another function of fire element is *nayana* or vision: being able to see forms. So now, in the unfolding flow of creation, forms solidify from airy and fluid movements into plasmic forms, further continuing the flows of air and water into more solid forms.

In the *Mahābhārata Śānti Parva* (227:52) Vahni is a *lokapāla* or guardian of fire, and is the mystic or secret name of *R – rakara*, the seed syllable for fire element as mentioned in the

Sutra. In Saivite philosophy and lore, Vahni is the fourth of the eight *Mahāmātṛs* or Great Measures of Creation (RAt is the fourth syllable in HaYaWa RAt) that unfold the universe. Vahni Mahāmātṛ is governed by *Unmatta Bhairava*, a fiery and powerful form of Shiva in the *Kubjikāmatatantra*. In another story, the Creator Brahmā assigned Vahni to protect the altar of the divine playhouse or theatre of creation.

Fire transmutes form into light so the soul can reveal its radiance.

Fire transports substances from the physical world into the non-physical world, and brings substances from the non-physical world into the physical world. In Vedic rituals these substances were offerings, be it ghee, flowers, prayers, intentions, mantras, and parts of ourselves we wish to change and dissolve.

Fire warms, inspires, ignites purpose, creates, dissolves, transforms and brings new energy. It is said that "fire never lies",[116] as fire is the only element that is not polluted by what it purifies. Fire is a generator and manifestor, a spark, an inspiration, a catalyst and igniter of decisive, bold, action.[117] Fire is forward movement – the urger on.

Fire is the light in life, the glow of the aura ignited, the shining spark in the eye, the glow of lustre and vitality. Fire is in the breath and the winds of force, fuelling the atomic matrix, the ATP in your cells that promote growth and give us our basic energy.

In *Yoga Vasistha*, Rishi Vasistha, Guru of Patanjali and Rama, teaches Lord Rama about the fire element:

Then I became the fire element through Teja – Dharana – concentrating on the fire element. Fire or light is predominantly sattva, therefore it is always luminous and dispels darkness even as a king makes thieves flee his presence. I realised the misery of darkness that destroys all good qualities because I became the light in which everything is seen.

Light bestows form on everything even as a father bestows form on his offspring. In the nether world light shines at a minimum level and there is greater darkness. In heaven, there is light alone and always. Light is the sun that makes the lotus of action blossom.

I became the good colour in gold, I became vitality and valour in men, in jewels I sparkled as their fire, and in rainclouds I became the light of the lightning. In passionate women, I became the twinkle of their eyes. I became the strength of the lion. I was myself the hatred of the demons in the gods and the hatred of the gods in the demons.

I became the vital essence of all beings. I experienced being the sun, the moon, the stars, precious stones, fire, lightning and lamp. When I became fire, the burning cinders became my teeth, smoke my hair and fuel was my food.

Sounding RAt

Vahni is a form of fire held in the sacral, solar plexus and heart chakras, and is known as a purifier and dispeller of darkness, a destroyer of disease and disharmony.

RAt is the seed syllable for Vahni Fire Element. *R* is a cerebral sound and a semi vowel, linking the vowels of light with the form making power of consonants. As you sound *R*, contract the navel and manipura solar plexus chakras, and flick the tongue against the roof of the mouth. Then, open the mouth wide in *AH,* slightly contracting the breath and mouth to end with a slightly sharp *t* with the tongue between the teeth, like a punctuation power point. *RAAt!*

Inhale RAAt into the navel chakra for 9, with attention here.

Now, exhale *RAAAT* out from the navel chakra for 9.

Repeat 9 times.

Now, do this with the solar plexus chakra,

and the heart chakra as well.

Repeat 9 times for each chakra.

176

Fire Meditation

The following meditation is adapted from Shiva's teachings to Shakti found in the *Vigyan Bhairava Tantra*, v. 29-30.

Take 6 deep breaths into your manipura solar plexus chakra and settle into this chakra of fire.

See your right big toe in flames whilst saying RAAt
See your next toe in flames whilst saying RAAt
See your next toe in flames whilst saying RAAt
See your next toe in flames whilst saying RAAt
See your right little toe in flames whilst saying RAAt
See your right ankle in flames whilst saying RAAt
See your right knee in flames whilst saying RAAt
See your right hip in flames whilst saying RAAt
See your navel chakra in flames whilst saying RAAt
See your solar plexus chakra in flames whilst saying RAAt
See your heart in flames whilst saying RAAt
See your right shoulder joint in flames whilst saying RAAt
See your right elbow in flames whilst saying RAAt
See your right wrist in flames whilst saying RAAt
Visualise your right hand in flames whilst saying RAAt
End at the solar plexus chakra, seeing it in flames.
Now, repeat for the left side of your body.
See your left big toe in flames whilst saying RAAt
See your next toe in flames whilst saying RAAt
See your next toe in flames whilst saying RAAt
See your next toe in flames whilst saying RAAt
See your left little toe in flames whilst saying RAAt
See your left ankle in flames whilst saying RAAt
See your left knee in flames whilst saying RAAt
See your left hip in flames whilst saying RAAt
See your navel chakra in flames whilst saying RAAt
See your solar plexus chakra in flames whilst saying RAAt
See your heart in flames whilst saying RAAt

See your left shoulder joint in flames whilst saying RAAt
See your left elbow in flames whilst saying RAAt
See your left wrist in flames whilst saying RAAt
See your left hand in flames whilst saying RAAt
End at the solar plexus chakra, seeing it in flames.
NOW:
See your whole body immersed in flame whilst saying RAAt
See your whole auric field immersed in flame saying RAAt
Now, see both bodies immersed in flames.
Say: "The universal fire flows through me."
Step into this fire totally, with all your heart.
Now, see the entire world consumed by flames.
Stay steady in this: do not waver.
Only not self can burn away.
Your essence renews in flame, for it is flame since the first moment of creation.

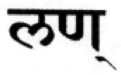

LAN: Earth

आधारभूतं भूतानामन्नादीनाम् च कारणम्।
अन्नाद्रेतस्ततो जीव: कारणत्वात् लण् ईरतिम् ॥१७॥

ādhārabhūtaṁ bhūtānāmannādīnāṁ cha kāraṇaṁ
annādrētastatō jīvaḥ kāraṇatvāt laṇ iritaṁ
17

Earth is the source for food.
From food develops the seed of semen,
and from that, embodiment occurs for the individual soul.
The earth is the basic source and support for all this.

Because of its importance,
it is said that from the letter *LA* evolves the earth.
LAn is the foundation for all elements,
nourishing all elements.
LAn is the foundational supportive element
for all embodied beings.

Earth is the glorious end of one cycle where the absolute stands as the glorious beginning on the other side. Earth is glorious because it is only through earth that all things of the world are created.
– Shankaracharya Saraswati

Earth Element is the sixth class of sound in the Maheshwara Sutra, and is connected to the element carbon which is number 6 in the Periodic Table. Uniquely amongst all five elements, Earth has its own class of sound in the Sutra because it is the culmination of the elements' alchemical processes, the landing ground for the elemental intelligences to coalesce into third-dimensional form. As such, Earth holds the blueprint of all elements within it.

Earth is the causal food for substance, nourishing the operating of all elements and the purpose of all elements. Earth is the food for our actions as individual selves. Earth is our supporter, nourisher and caregiver, the cause for living beings, the ground on which we exist, the source for food and growth. From food arises semen, reproducing and continuing our species, proving that earth feeds all substance. Without *Anna* or food, our bodies would not exist.

Earth is our foundation for expansion, supporting, upholding, grounding, stabilizing and completing. The more grounded we are, the higher we can fly. Earth forms our limits and boundaries, enabling us to define ourselves in order to fully express and manifest ourselves. Earth anchors self, creating a container for our lives, establishing healthy boundaries to feel whole within ourselves.

In earth element, we find the things that matter to us, that which has value and meaning. Earth is Manifestation, the stability to maintain awareness and ground purpose in direction. Conversely, in earth element we meet the fear of being stuck, unable to move, feeling trapped and limited to our bodies.

In embracing this fear, Earth allows us to settle, root, and find our ground. The physical body is coagulated into form by earth element, held in the blueprint of Earth's frequency. Being defined with a body allows us to anchor and experience all vibrations into the physical here and now: to embody. Joy and freedom expresses through us from *A I Un* as it manifests in our bodies through *LAn*.

The Earth Frequency

Truth, high and potent Law, the Consecrating Rite, Fervour, Brahma and Sacrifice uphold the Earth. May she, the Queen of all that is, and is to be, may Prithivī make ample space and room for us. She who is Lady of the earth's four regions, in whom our food and corn-lands have their being, nurse in each place of breathing, moving creatures; this Earth vouch safe us kin with milk that fails not.
– Atharva Veda[118] 12:1-63, Hymn to Prithvi

Life on earth is assumed to have begun in a mix of aqueous and oily compounds, sunlight and lightning, called the "Primordial soup", as theorized by Oparin and Haldane in the early 20th century. This primordial soup of the elements as seen in the previous class of sound HAYAWARAt now coalesces into matter in LAN.

In the 1950s Miller and Urey simulated a version of the primordial soup and found "amphipathic" biomolecules such as tryptophan and, most interestingly, psychoactive molecules like the feel-good chemicals of dopamine and serotonin, and the mind altering substances of LSD and DMT that can give one insights and experiences into the nature of creation.

The Earth frequency has hovered for many years around 7.8Hz until 2015, when it started to fluctuate. This frequency, known as the Schumann Resonance, is a vibrating wave-field or resonant hum of the Earth's vibration that is the foundational supporting ground frequency for all living organisms on earth.

Whenever your frequency is out of alignment with Earth's frequency, you become out of sync and out of alignment with your own natural rhythms, physically, emotionally, mentally and spiritually. As Earth's frequency is now shifting and changing, this causes many disturbances, diseases and stresses for us because our electromagnetic fields and the resonant frequency of Earth are symbiotically, inextricably, intimately intertwined.

Many indigenous societies have maintained a living connection to the earth frequency as part of their daily way of life, which is why they move more slowly and are generally happier than their "first world" counterparts.

Living in resonance with earth's frequency slows down the experience of time for us. We align into a relaxed, patient, peaceful and content way of life. The more we connect with Earth frequency, the more emotionally attuned we become, able to flow, let go into, and trust the natural feelings and cycles of life.

The more we connect with earth frequency, the higher our own vibratory rate becomes. The higher your vibration is, the stiller you become. Earth helps us achieve this by helping us slow down and become more still, changing our experience of time.

If we look at the Aboriginals, they measure time in millennia, not days or months. They are in no hurry at all, as they live a life-affirming wisdom that comes from moving slowly, in tune and at ease with the cycles of life and death.

Our DNA connects to Earth and its electromagnetic fields. This resonance is part of the foundation for the Law of Attraction – the universal law that draws towards you that which serves your purpose and learning, be it relationships, teachings, events or resources. Until you resonate fully with the Earth, this law is not fully realized in your life. Earth is part of the attractor, part of the gravity of this law.

The memories of everything that has ever happened on earth are stored in her magnetic fields, just as our own memories are connected to our own magnetic fields. Everything that has happened to us on earth, in all our incarnations, are stored in earth's fields and within our magnetic fields. We are connected to, and mirrored by, the electromagnetic fields of Earth. In the concept of reincarnation, we only return to earth because we have unresolved learnings here, all of which are stored in the earth's fields and our own DNA. Earth is a powerful attractor!

Many of our emotional memories are connected to, and held within, Earth's magnetic fields. NASA astronauts found that when they first left the earth's orbit and magnetic fields, they lost connection to some of their memories and identity and became disorientated. Now every astronaut has a magnetic attenuation box on their belt so they do not lose their emotional memories, reference points and sense of orientation.

Similarly, NASA astronaut Scott Kelly found that spending nearly a year in space away from the earth frequency significantly changed his DNA. "Scott's telomeres (end caps of chromosomes that shorten as one ages) became significantly longer in space."[119]

He had hundreds of "space genes" activated by the yearlong flight away from earth that altered his "immune system, DNA repair, bone formation networks and hypoxia, resulting in him growing two inches in height." NASA confirmed that 7% of his genes remain changed and they may stay this way because of "space travel, which causes changes in a cell's biological pathways and ejection of DNA and RNA."

Similarly, researcher Valerie Hunt did experiments with people in special rooms (on earth!) where they removed the background Earth field and frequency. People became deeply emotional. When the Earth field was restored, people felt balanced and "normal" again.

An experiment in 2011 by Nobel Prize winner Dr Luc Montagnier further shows us the effect the earth's fields have on us. He removed DNA from test tubes filled with water samples and exposed them to the Earth frequency. The water, which had no DNA, *produced new molecules even though no life was present.* He theorized that our DNA communicates via Earth's electromagnetic field, which we are all connected to. The frequency of the Earth not only sustains life, but *participates in creating it too.*

The Seven Dhatus

Lan is the cause of food, from where semen is created, and from this, the individual self is birthed.

In Ayurveda there are seven *dhatus* or building blocks of the body. The first is *anna*, the food we eat. Food is converted through chewing into *rasa*, a liquid that nourishes all the tissues, organs, and systems of the body to form *rakta*, red blood cells that oxygenate tissues and organs.

This transforms into muscle or *mamsa*, and then *meda*, or fat, which maintains the lubrication and operation of the body. This transforms into lymph and *ashti*, or bone, for the skeletal frame. Bone further refines into *majja* bone marrow, a concentrated form of energy, with the final step being the production of *sukra* semen, the most concentrated, potent, fluid force that generates the individual self.

Lan is the cause of all of this. Without food there is no semen, and no way to reproduce the species. As Lan is the cause, so it is the beginning of our earth life and human form.

Ways to Commune with Earth Element

An awakening must come in the earth nature and the earth consciousness, which will be, if not the actual beginning, at least the effective preparation and the first steps of its evolution toward a new and diviner world order.
– Sri Aurobindo

1. Walk barefoot on the earth. Sit and lie on the earth every day.
2. Breathe through your heart, womb/hara and feet down into Earth's Core and back up from her Core through your feet, womb/hara to rest in heart.
3. Work with Earth's ley lines in sacred sites.
4. Wear, meditate with, and use Turquoise and Nephrite Jade.
5. Use the 6 Earth Sounds of the Sutra and the 5 root chakra sounds.

Earth and Heart Meditation

Feel, from your heart, a beautiful place in nature that is **your sacred site and power spot**, the place in nature *you most resonate* with: a river, a park, a forest, a sacred site, the one that comes to you immediately. Feel/See the details: the beach, the ocean, trees, animals, birds, scenery and energy of the place itself.

Feel the love, gratitude, appreciation you have for Mother Earth. Say: "Beloved Mother Earth, *I love you, I love you!*" Once you feel this, send your love down into the Womb of Earth, and say: "Beloved Mother Earth, please send me your Love." Wait for Mother Earth to send her love back to you. When you feel this love, let it move throughout your heart, your soul, your womb, and your body.[120]

Sounding Earth Element

LAN is a dental sound. The tongue lightly touches the tip of the upper teeth with *L*. Open the mouth wide with *LA*, with long *AH*. The *NN* can be sounded into the pineal directly through the roof of the palate. You can also sound *LAAN* directly into the muladhara root chakra at the base of the spine. *See which one works best for you.*

ELEMENTAL SUPERSTRINGS OF SOUND

The 42 Vibrational Medicines
of the Maheshwara Sutra

The five elements compose the energy and matter for all physical creation. They deeply influence and affect our bodies and the state of our mental and emotional health. Ayurveda uses the five elements to both diagnose and heal patients, and originally used mantras composed of the sounds of the elements to heal patients. Mantras were prescribed and dispensed, rather than just pills, powders and potions.

Ayurveda was originally a science of mantras concocted by Patanjali, one of the sages present to receive the vibrational medicines and sonic remedies of the Maheshwara Sutra. From the sonic teaching delivered by Maheshwara, Patanjali brought forth vibrational medicine mantras.

The 42 emanations of creation in the Maheshwara Sutra are 42 vibrations, and when we identify which ones are blocked in us, which vibrations we have fallen out of tune with, we can use the appropriate vibrations and sounds to "re-tune" us.

The 42 sounds are a vibrational pharmacy, a vibrational medicine kit of frequencies we can use to come into tune with our harmonious blueprint, free of distortions. The more we tune into these 42 vibrational medicine frequencies, the more we tune into our original harmonious blueprint composed of the 42 aspects of the *Aham* or I AM identity self.

As we saw in verse 4 of the Sutra, *"With the union of the first and last letters of the Sutra. A and Ha, is Aham, the I Am,"* or we could say that *"the existent Self AHAM is generated in all its detail through all the sounds, from the first to the last."*

This is a simply put revelation: the 42 sounds of the Maheshwara Sutra are what the universe is made of *and* what the "I" is made of. The sequence of sounds from *A to HA* contains the full range of experiences available to your individuated self. When these sound frequencies are sounded correctly, you entrain and attune to them.

Vibrational waves composed of multiple layers of frequencies are resonating, vibrating and passing through our electromagnetic fields, nervous systems, cells and organs *all the time*. These can be translated as sounds, and these sounds are what the Maheshwara Sutra is composed of.

In sound healing terms, the frequencies and sounds that are muffled, blocked or lacking within you need to be sounded and inputted into you, so you can receive them and entrain to them. Some parts of us may be missing these frequencies, or maybe we are too stressed to access them. Sound healing focuses on bringing these frequencies to us, like a form of nutrition. Just as we have dieticians and nutritionists, so we now have vibrational pharmacists.

Over time, the body will respond and produce the frequencies it has been lacking, or needing more of, by itself. Vibrational medicine can trigger the body-mind into activating its own medicine; then the external sounds are no longer needed.

All frequencies resonate within different subtle or gross forms of matter. We can call these different substances. Our thoughts and emotions vibrate in different substances: they feel different. Our emotions are felt in one substance, our thoughts are in another substance.

When we vibrate these substances, we live in with specific intentions and sounds, we can change them, and change ourselves. If there are holes or blocks or stagnant, nullified areas in these substances, where the substance is incoherent and not flowing, we can bring the flow back.

Sound and frequency is intelligent. We can apply them specifically. For example, if you need more water, more emotional healing, more release, sound *VA* and its associated sounds. If you need more fire, energy, catalysing, action and power, sound *RAt* and its associated syllables.

There are many combinations of sound healing codes in the Sutra. Combinations can be the most powerful medicines. Throughout the Sutra, most of the sounds that arise after the elemental sounds *HAYAWARAT-LAN* correspond to one of the five elements. So, when we thread together, like pearls on a string, the appropriate sounds for each of the five elements, it makes an elemental *superstring of sound*.

These elemental superstrings of sound show how the elements vibrate and work within the different parts of us. By making these sounds we can attune to the harmonious flow of the elements within us.

Additionally, each individual has their own unique vibrational signature, which is often muffled or blocked by our conditioning and wounds. To access your own unique vibrational signature one has to release the blocks in its way, which you can do through creating your own superstring vibrational remedy tailor-made for you.

This mantric prescription is your own vibrational medicine remedy, which can be from 3 to 11 sounds long, depending on the individual. The length of time for doing your mantric remedy will vary from person to person. Additionally, you can also discover your own vibrational signature by using combinations of the sounds in the Sutra.

Elemental Superstrings

Each of the five elements has five to six functions and sounds within us which also course throughout creation. These sounds are given throughout the Sutra, and they weave into superstrings of vibration which are composed of:

1. The individual element or *mahabhuta*
2. The 5 elemental measures or *tanmatras*
3. The 5 functions of Speech, Handling, Movement, Excretion, Procreation: the *Karmendriyas* or organs of action
4. The 5 sense organs or *Jnanendriyas*
5. The 5 *Pranas* and 3 aspects of mind
6. The 3 *gunas*

Linking all the sounds of the five elements together in vibrational medicine mantras can help tune one into the underlying harmonious flow of each element throughout our body-mind-spirit, as well as help heal disorders connected to them.

We all have strengths with certain elements, and certain weaknesses in other elements, and we need to balance this out. For example, you may identify more as a watery person, and not have much fire. You may be a more airy person who needs more earth grounding element. Balance is key. We need to give ourselves what we need, once we identify what is lacking within us.

In sounding these elemental mantric superstrings, use the *toroidal wave-flow order of the chakras* as given by Shiva in the Sutra, and resonate each chakra with each superstring 11 times each.

1. Pineal
2. Throat
3. Heart
4. Sacral
5. Solar Plexus
6. Root Chakra

All the seed sounds of the elements and their five or six associated functions can be pronounced with a long AH and with anusvara MM at the end so they become, for example, HAAMM, YAAAM, WAAAM, RAt or RAAMM and LAAAN.

Alternatively, you can use the seed syllables without the MM. If you do use MM, remember to vibrate MM nasally and into the roof of the palate and the pineal, *not vibrating the lips*. It may help to keep the mouth open slightly to do this. There is also a mudra for each element for you to hold, which connects the relevant electromagnetic circuit within you, whilst doing the sounds. The orders of the sounds can be done in different ways: linear as given in the Sutra, or in sequences that suit your unique disposition. Experiment and discover what works for you.

The 6 to 9 sounds that compose each Elemental Superstring Vibrational Remedy are shared in depth and detail in the Shiva's Hologram Courses, Workshops and Retreats. Each of these individual Elemental Superstrings flows through, and creates, the body-mind-spirit system.

NA MA NA NA NAM:
Measuring the Elements

शब्दस्पर्शौ रूपरसगन्धाश्च अमङणनम्।
व्योमादीनां गुणा ह्येते जानीयात् सर्ववस्तुषु॥१८ ॥

Sabdasparśau rūparasagandhāścha ñamaṅaṇanaṁ
vyōmādīnāṁ guṇā hyētē jānīyāt sarvavastuṣū
18

Sound, touch, form, taste, smell are *namanananam*,
found in the generation of all things.
From space these threads evolve,
arising from eternal substance.

"What is the soul, bound by light and dark?" Brahma-Prajapati answered, *"There are five subtle elements tanmatras... there are also five gross elements mahabhutas. The Union of these is the human body."*[121]

Namananaṁ are the five *tanmatras* or "measures of that", which unfold, explicate and express the five elements in the body-mind system.

Tan means to extend, propagate, continue, stretch a cord, spread, spin out and weave. By doing so it measures *matra* out each element into manifestation, creating a *"pathway of measures"*. *Tan* is the extending and unpacking out of each element in toroidal spirals to weave the sensations and capacities for sound, touch, form, seeing, taste and smell throughout the body-mind system.

Tan also means to render, to *cause something to become usable*, and in this context, relates to our ability to see forms. In modern terms, software graphics programs calculate the appearance of objects and how to use light and shadow (polarities) in order to make objects look realistic on a screen. This then enables us to construct an artificial world that looks real, such as we find in movies, video games, CGI special effects and architectural designs.

All these CGI designs are "rendered" over time with many computers in "rendering farms", as huge amounts of computing power (the equivalent is the data processing power in our brains) are necessary to render the special effects of our favourite movies and video games, thus creating the appearance of objects through using light and shadow (polarities). This of course is similar to how the brain works as well, creating our own virtual reality simulation.

Tan also means "to prepare a way for". In this case, *tanmatra* is preparing the way for the further expression of each primary element through its *matra* or measure. Like a ball of string, the elements' threads are only seen when it is unfurled, and this is its matra. In Patanjali's *Mahābhāṣya*, matra is defined as "separation, splitting up a compound word (in this context the four elements of HaYaWaRat) into its constituent parts."[122]

Matras are periods of time *and* spaces of silence that measure boundaries, and boundaries create edges: the forms of things. The spaces or matras between sounds are as important as the sounds themselves, as these spaces allow the interacting elements to express and interact, imprinting the shapes of their dance into their substances, thus creating moulds of subtle forms.

When Sanskrit words are sounded, as in the *Pada Patha* of the Vedic *Samhitas*, there is a momentary pause measuring one matra or time-period required to say a short vowel.[123] This pause allows the sound to vibrate the substance, creating a subtle shape that then guides the next movement of vibration. Without this matra or pause, the new shape or form cannot occur.

The space between sounds, and the silence after sounding a mantra, allows the energy generated by the sound to be fully felt and received, deepening its imprint within you and bringing it to its fullness. Receiving the sound is as important as generating it, for in receiving it you access its full spectrum of vibratory nuances, allowing it to permeate you more deeply and imprint you more thoroughly with its full power, which is found in its subtlety and the space it takes you into.

When you sound mantra, one needs the right spacing when doing it, just as meditation is most effective when you rest in the spaces between your thoughts. In these *moments of measure*, of sound and silence, rhythm and space, the tanmatras reveal.

As Rishi Nandikeshvara says, the tanmatras are *"found in the generation of all created things."* Without the tanmatras, the five elements would be a swirling soup of energies with no measures or further expressions of them to become usable for us. Thus, for each element to become usable for us, it has to have its measure.[124]

Traditionally, each of the five tanmatras are the home or container for the sensations of each of the five senses.[125] For example, whatever we say is stored in *shabda tanmatra*, the home for the impression of sounds. The sound remains here, and it can be retrieved and created again from this container. This works the same for each of the five tanmatras and five senses.

The five tanmatras arise from space element, all linked through the Phi Golden Mean spiral, which explicates and expands its code from the five elements into the tanmatras. In the curvature of space-time and its toroidal vortex spirals, the tanmatras are both the five elements' offspring *and* their generation into form, cause and effect interweaving and exchanging with and in each other. This creative process works both ways, each feeding the other in toroidal loops, continuously manifesting sensory and supra-sensory perceptions and ways of enjoying reality.

Tanmatra Tantra

The Tanmatras are our *subtle sense perceptions* and doors to sensations and impressions, creating the rich potential for experiences in sound, touch, sight, taste and smell. When we tune into these subtle senses, they can expand your perception of the world, making you more engaged and alive in your environment and with the beings around you.

The tanmatras are not just the basis for these heightened senses; they are the origin of our "super senses". Just as the physical senses help us navigate and experience earthly life, the super senses open us to the supra-sensory world helping us touch, see, hear, taste and smell multidimensionally in ways that are beautiful, life enhancing, joy inducing and gratitude filling.

Each super sense deepens our experience of the sensual texture of life and the supra perceptive faculties. Tantric practices work with the tanmatras as vehicles of the quantum code in order to enjoy the infinite in many ways.

One way is through the supersensory and multidimensional perceptions of clairvoyance/fire element, seeing things beyond the visual range, clairaudience/space element, hearing beyond the auditory range, clairsentience/air element or touching invisible energy fields, clairalience/earth element or smelling fragrances beyond our world, clairgustance/water element or tasting beyond our world. We can perceive through these supersensory perceptions when the mirror of the mind is clear enough.

The tanmatras can be experienced as delight in our senses and our super senses when One sees and enjoys forms, hears sounds that arouse, drinks in refined, sublime tastes, caresses exquisite sensations of touch, and are intoxicated on the nectar of fragrances beyond this world. This enjoyment is also the extending of the One's essence, which does not grasp onto or identify with any of these sensations or experiences. One simply enjoys it as a play of creation, and lets the sensation go.

The first tanmatra is sound *shabda* arising from space *vyoman*. Through sound you can change the substance of space by vibrating it, within you and around you. All tanmatras vibrate in a substance, be it space, air, water, fire or earth, and this differentiates all the tanmatras.

What substance you make a sound in will determine what effect the sound has. The manner in which sound vibrates in any substance determines what form the sound will create. The substance also depends on which of the four modes of sound your consciousness is *in at that moment*, as discussed previously.

The second tanmatra is touch: sound carried from space through air. Air touches, connects and moves, but the root cause of touch is sound. If you listen to air rustling leaves, if you listen to your own breathing, the waves of the ocean, all air has sound.

The third tanmatra of fire is vision and form, which has sound, touch and now form. The invisible now becomes visible. The fourth tanmatra of water has sound, touch, is visible and now has taste. The fifth tanmatra earth is smell, which completes the chain of manifestation as earth element solidifies all tanmatras.

1. Space   Sound
2. Wind   Sound + touch
3. Fire   Sound + touch + form
4. Water   Sound + touch + form + taste
5. Earth   Sound + touch + form + taste + smell

Element	Matter	Occurrence	Sense	Organs	Consciousness
Earth	Solid	Shadow	Smell	Organs of Reproduction	Stabilizing
Water	Liquid	Reflected Liquid Light	Taste	Organs of Evacuation	Nourishing
Fire	Plasma	Heat Sound Light	Sight	Eyes	Vision
Air	Gas	Electrical Energy	Touch	Hands	Movement
Space	Etheric	Wave Field	Sound	Ears + Voice	Medium for manifesting

Space element is the medium for all things to manifest, from which comes sound, the voice and hearing. With air comes touch, handling and movement; with water comes taste, enjoyment, emotion, release, nourishing; with fire comes sight, form and light; with earth comes smell, solidity, stabilizing and culmination of all elements.

All five elements spin in the vortex of resonance *NAMANANANAM*, where the tanmatras act as measures and bridges to manifestation which we perceive through our super senses, the origin of the 3rd dimensional five senses.

NA: Sound – Shabda Tanmatra

The word *guna* is used in this verse to express the unfolding of each element's attributes, qualities and values. A *guna* is an ingredient, constituent, thread, a string of a musical instrument, and a subdivision measuring out part of a whole. Guna is an action of doing, repeating, dividing and multiplying itself. A guna is also esoterically seen as a specific property of letters, as seen in *ñamanananaṁ*.

Each of the five tanmatras are the home container for the sensations of the five senses.[126] When you speak a sound, this sound goes into its home of sound-tanmatra or *shabda tanmatra*. Whatever we say is stored in shabda tanmatra, the home for the impression of sounds. The sound remains in shabda tanmatra and it can be created again from this tanmatra. This is the same for each of the tanmatras and senses.

The measure or expression of the intelligent awareness of space is *shabda* sound. Self manifests through sound in the wave-medium of space. In the *Purva Mimamsa* by *Jaimini*, "sabda arises from the origin, and the knowledge revealed by this is the original teaching and authentic means of right understanding."

Sabda is self-evident, relying on nothing else for its proof, as to hear directly is to understand directly. Its true meaning is not created by human reasoning but is rather revealed to human understanding. Any sound that is made audible by pronouncing it only shows that the sound exists *before* pronouncing it: it is already here. Pronouncing it brings the sound into our dimension.

It may appear sound comes and goes, but this is only because contact with the object is lost. When we hear a sound, we usually do not listen to the source of the sound. If we deeply listen to and follow the sound we hear or make into its source of space, we access shabda. Shabda tanmatra holds the impression of the sound, and we can access this. In this, sound and space are one.

In subtle hearing beyond the audible range known as *clairaudience*, you can hear intuitive information from your Self, non-physical guides and other non-physical beings *internally*. In clairaudience we can receive these messages and hear these voices, which might be in our language or others we do not intellectually know. Many spirits can communicate to you, and these voices may sound different depending on the spirit communicating to you – the voice may sound as if it is right next to you, or coming from far away.

In clairaudience one can hear non-physical intelligences helping you, and negative spirits trying to influence you. The voice may give you advice to uplift you, or distract you. One may also hear the voices of your ancestors, nature spirits, animals, devas, gods and goddesses.[127]

All beings communicate, and most of them do so through non-audible means. Many beings are ready and willing to dialogue with you, and all one has to do is reach out to them through your focused thoughts to initiate this dialogue.

In clairaudience, one can also hear forms of music and choirs, buzzing sounds, bells and a vast range of other sounds that demark certain states of consciousness, as the yogic texts tell us. Hearing the subtle sounds of creation are an integral part of yogic practice.

From the sound of bees to the sound of AUM, yogis have heard celestial and elemental sounds for millennia. In the most refined state of clairaudience, one can hear-feel the vibration of objects and people. As objects and vibrations are the same in *nama-rupa*, hearing-feeling-seeing these allows one to penetrate an object and alter it according to your will, leading to phenomena such as teleportation and changing the molecular structure and elemental composition of objects: changing the proportion of water in an object into fire for example.

Perhaps the most useful subtle sound is the sound of your own soul, the still, small voice within you. To hear the voice of your soul means you can reunite all parts of yourself around this one voice that is part of the quantum code. In this gathering, we return to the field that flows throughout all creation: space.

Telepathy is conscious communication in the clairaudient state: space and shabda tanmatra. Receiving this form of sound from a person who is consciously in space element is a powerful transmission.

Sounding NA

All the tanmatras are nasal in quality, resonating the pineal. *NA* is a palatal nasal sound. With mouth open, tongue touching the top of the lower teeth, *NA* is resonated in the upper nostril bridge and pineal. If you pinch the nostrils shut, the sound ceases.

Deep Listening Practice

Part of the power of sound comes from listening deeply. As Yogi Bhajan shares, "Man is born... to listen. This is one secret which, if you will never learn, you will never have the essence of life. When you stop listening, you don't want to listen. That is the time when you make the greatest mistake, because your constitution, your construction, your building, your faculty, your power, is in listening."

One can hear and feel the sound frequency of people through their voices, which contain the sonic signature of their soul. To do this, we have to listen. Real listening is a selfless presence. As Krishnamurti shares,[128] "the very act of listening brings its own freedom."

Deep listening attunes one to the subtle element of sound or *shabda tanmatra*. Distinguished sound pioneer Alfred Tomatis believed that with proper listening to sound, we can establish communication with the cosmos. In the Kundalini Yoga practice below, a transference of healing energy can happen, bringing both people into a deep, clear peaceful space. All through listening!

Do this *shabda* or sound healing slowly and mindfully. Start by sitting up next to a partner, with your partner lying down. Start with closing your eyes and bringing your awareness to yourself, centring yourself. Take a few deep breaths into your heart, arriving and resting here, bringing your full presence here.

Now gently focus your awareness on your listening.

Listen to the breath of your partner.

Listen to the heartbeat of your partner.

Become aware of some more of your partner's body sounds.

Listen to their cells.

Listen to the sounds of life in your partner's body.

Listen to the stillness between the sounds, the silences surrounding the sounds.

Now, listen to the sounds in the surrounding room.

Listen further now to the sounds outside the room.

Listen to the birds in the trees, the birds in the sky, the gentle wind, any other noises.

Allow your hearing to expand further outside into nature and Mother Earth sounds, on the land, in the water, in the sky.

Listen to the drums in Africa,

the sound of a tiger in the jungle in Asia,

whales singing in the ocean in the Pacific.

Listen further and further.

Listen to all the sounds on this planet.

Then, extending even further:

Listen to the sounds of light, of the Sun. Listen to the Cosmos, the close by and distant stars and Milky Way: Receive the sounds of light as you listen.

Now, follow this listening thread to the source of all these sounds.

Now listen to the Cosmos again, the far and distant stars, the Milky Way, the stars closer by, the sun and moon, the sounds of planet Earth – the sounds in the desert, the oceans, and the forests.

Listen to the sounds back closer to home now: in the backyard, the birds in the trees and children playing.

Listen to the sounds in the room, the breath and the heartbeat of your partner. Listen to your own breath and heartbeat.

Allow yourself to rest in this space.

Be Present here now.

MA: Touch – Vayu Tanmatra

The sense of touch pervades all the senses. It is permanently associated with the mind. The mind pervades the sense of touch. This in turn pervades all the senses.
– Caraka Samhit Sutrasthana 11:38

Vayu Air element of energy in movement that manifests the Self has the tanmatra of *sparsha* – touch. Anything that touches something else connects, feels and relates with it, and in so doing creates an experience. The nature of experience is ever moving, created by the touch of one thing to another. Touch is part of the mind, which pervades the sense of touch.[129] Mind is associated with Air element.

Touch is a fundamental need of life. We all need to touch, and be touched. We all need affection and contact, to love and be loved. Through touch we feel alive and connected, and can transmit ourself to others, exchanging energy and information on many levels.

As air arises from space to carry movement, touch arises from sound to feel our vibrating, moving connection with life. Waves of information come from touching and being touched, softening and opening us up to more and new possibilities, new potentials and new ways of thinking, feeling and acting. We accelerate our growth through touch exchange, accessing more than what we can do alone without touch.

In her book *The Power of Touch* Phyllis Davis describes how touch is an actual biological need, just as vital as food and sleep is for our emotional, physical and mental health. We express our spirit through touch, because touch is the first sense we physically experience in the womb. We grow in constant physical touch with our mother, and touch is our first memory of nurturance, safety, security and comfort.

This "skin hunger" is as real as food hunger. When consistent care and touch is not given to a child between 9-27 months, their amygdala and hippocampus do not grow to their full emotional capacity, as these brain centres are responsible for pruning neurons for emotional bonding.[130]

Without touch at this crucial time in our development, we lack emotional bonding, resulting in unhealthy attachment and psychological issues in the child's later life, which can further manifest as codependence, coldness/isolation in relationships, and sexual intimacy issues.

Cultures in which infants are given lots of touch have lower incidences of theft, murder and rape. In experiments with baby monkeys, researchers shared two fake metal "mothers" to them. One was hard and cold, but dispensed milk. The other was soft to the touch and comforting, but provided no milk. The babies went for the soft touch, showing that the need for touch is as important as the need for physical survival.

Touch is an essential human need, life affirming to our bodies, our minds and our very sense of feeling human. The absence of touch can lead to alienation, isolation, loneliness and disconnection from parts of oneself and others.

Touch brings us into places where we risk being open and vulnerable. Allowing oneself to be touched by suffering *and* by love, to touch our own emotions and allow them to move us, lets life refresh us. Feeling pleasure and pain, to be touched by joy and suffering, empathy and compassion, is critical to our life, and the more we do this the more we feel how others feel, developing our own senses of empathy and compassion.

Touch stimulates self-awareness beyond the physical into the emotional. When we hug someone authentically and heartfully, the body releases oxytocin and serotonin which makes us feel good, reducing the harmful impact of stress on your blood pressure and heart rate, lowering stress hormones like cortisol.

Neuro-economist Paul Zak recommends at least eight hugs a day to be happier and enjoy better relationships. Shekar Raman MD in *The Huffington Post* shares that: "A hug, pat on the back, and even a friendly handshake are processed by the reward centre in the central nervous system, which is why they can have a powerful impact on the human psyche, making us feel happiness and joy.

And it doesn't matter if you're the toucher or touched. The more you connect with others, on even the smallest physical level, the happier you will be."

Touch and feeling is how we get feedback and emotionally engage with the world, and is an empathic, connecting, bonding sense. Both physical and energetic touch connects our sensory experiences with our emotions, connecting our multidimensional self to our body. Touch tells us intuitively what is real and gives the mind proof of what is real, for to touch something is to know that it is real.[131]

Clairsentience or "clear-feeling" is the ability to feel and touch subtle information that lies beyond the physical dimension. Being clairsentient allows you to feel the unique electromagnetic fields of other people, non-physical entities, and places.

Clairsentients feel the energy of their surroundings, describing places as clear, heavy, stuck or light. They can feel when there are ghosts or spirits, positive and negative, around them. In walking into a room, they can tell if people are relaxed or if there is tension, and they can feel how someone is feeling without them saying anything. A clairsentient can easily relate to others, understanding their emotions and where they are coming from without. They can read people and feel their pain and emotions, and can tell when someone is telling the truth or lying.

Clairsentients are empathic and can absorb the energies of other people, buildings and places. Their mood can fluctuate depending on who they have talked to, and what they are surrounded by. Another form of this is clairtangency (clear touching) or psychometry, where one can perceive the history of a place or person by touching them or an object they owned.

For example, if a psychometrist holds a person's watch or touches their hand, they can tell what has happened to them as well as the history of that object and where it has been. Similarly, if they walk into a building they can sense the history of the place, and what significant energetic events happened there.

Self-Inquiry

1. In your daily life, in what situations do you touch other people or experience them touching you? Who initiates touching? How important is touch for you in greeting someone? Is touching others something you welcome or something you try to avoid? When do you feel in need of touch? What happens when you are deprived of touch? *Giving and receiving touch are equally as beneficial. If you feel in need of touch, give it! It will make you feel better too.*

2. We heighten our sense of touch by closing our eyes, for example when we meditate or pray, so as to feel touched by something greater than ourselves.

Choose a small object that fits into your hand, which symbolises the sacred for you. Close your eyes and take a few minutes to feel into and explore this object. Slowly explore it with your fingers: the shape, texture, temperature and size. What words come to you that define how it feels? As you notice how it feels, in what ways does this mirror your experience and relationship with the sacred touching you, and you touching the sacred?[132]

Sounding MA

M is a labial and nasal sound. With tongue relaxed and lips pursed as if to blow out a candle, *MA* is directed from the bridge of the nostrils and pineal, and minimally vibrates the lips.

N: Form – Rupa Tanmatra

Of forms, the eye is the source, for all forms arise from it. It is their common feature, for it is common to all forms. It is their very Self, for it sustains all forms.
– Brhadaranyaka Upanishad 1:6.12

The tanmatra of fire element is *rupa*: form. Rupa is twinned with *nama*: the name or cluster of sound frequencies that are the sonic signature of an object. *Nama-rupa*, vibration-form, is the unique identity and fingerprint of a form. Every form arises from waves of vibration which create an endless array of shapes and structures.

Through the power of the Name or the Word, the Creator sees, names and therefore brings forth the form of each and every thing from within the quantum field. In the creation story of Brahma, he reveals the forms of the world by naming them through vibration, as did Adam. In Sumeria, sequences of sound or mantric like spells known as *"me"* were understood to form the Universe. In Palestine, the 215 Letter Name of YHWH was said to summon forth the Creator Itself.

In the *mahavakya* or great saying of the *Purva Mimamsa* *"mantratmako devah"* the gods exist in their sounds: their forms and qualities are made from mantra. "By mantra the gods cloak themselves, by mantra they are revealed." In the science of Cymatics, sound vibrations form distinct geometric shapes, as we saw earlier in the air and water elements.

The tanmatra of form arises with the other expressions of the fire element: the eyes and the power of vision. This was eloquently expressed in the "Fire Sermon" given by the Buddha on Vulture Mountain to 1000 priests, where he said:

All things are on fire. And what are all these things that are on fire? The eye is on fire; forms are on fire; eye-consciousness is on fire; impressions received by the eye are on fire; and whatever sensation, pleasant, unpleasant or indifferent, that originates in depending on impressions received by the eye, that also is on fire.

The fire element, seeing, and the eyes were understood by both Buddha and Christ. As Yeshua shares in Matthew 6, *"The light of the body is the eye, if therefore thine eye be single, the whole body shall be full of light."*

The eye is an expression of fire element, giving us the capacity to see forms (rupa) as fire element reveals forms.[133] The light of the body and all forms can be seen when your eye is single, one pointed and focused on a form.

Through this focus, you can also see beyond this light into the formless. We can use the phenomena of forms to access the formless. Visible forms are gateways into subtle forms and from there into the formless. Architects worldwide for millennia have utilized the science of sound to create forms in physical sacred spaces in order to enter this luminous world of subtle forms and the formless.

One example of this is Chartres Cathedral, France, a powerful sonic and geometric architecture created to resonate with the harmonic proportions of the human form. Entering Chartres is like walking into a world of many possibilities, where you immediately start to vibrationally entrain to your innately harmonious proportions.

Orbs of light hang suspended in darkness, and pristine geometries ripple in blood red and blue waves throughout the building. Throughout the day the light inside the cathedral constantly changes in intensity and colour. Certain windows are coded to catch the attention at certain times of the day, reminding one of the Indian Raga musical scale, whereby particular sets of notes are played at specific times such as morning, twilight, midnight and so on.

Each stained glass coloured window, like music, is played differently by sunlight at different times of the day, immersing the interior in a different atmosphere or vibratory rate. This provides us with a new facet, a new chord, and a new face of consciousness every hour of the day, with light, shade and colour subtly combined or dramatically contrasted.[134]

Chartres Cathedral is built to reflect this dance of light and shadow, *rendering* its subtle forms to help us see the beauty of material objects. It then uses this perception of beauty to go beyond the form itself, transporting our consciousness into a space that is material and spiritual *at the same time*. It uses the visible world as a gateway into the invisible world, just as many other scientifically designed sacred spaces do worldwide.

As Abbot Suger, one of the first Abbots of Chartres 700 years ago said, the *"dull mind rises to truth through that which is material and seeing this light, is resurrected from its former submersion."* We can use what we see as gateways to access non-physical worlds and their subtler forms. As Christ[135] shares, "Come ye from invisible things, to the end of those that are visible, and the very emanation of Thought will reveal to you how faith in those things that are not visible is found in those that are visible."

Rupa Tanmatra is also the visual perceiving of light, colour and form that can then be harnessed, refined and focused by meditating on forms such as sacred geometries, mandalas and yantras. Traditionally, the star tetrahedron is the geometry connected to fire and forms, and recent scientific and philosophical inquiry shows that tetrahedrons do indeed pattern the forms of our reality.

According to Dr David Frawley aka Pandit Vamadeva Shastri,

Colour (Rupa Tanmatra) is an important therapy (Tanmatra Chiktisa) in Yoga and Ayurveda. The right use of colour stimulates prana, harmonizes emotions, opens the inner eye, and connects us to the devotional light of the Deva Lokas. It is a connection to higher realms of awareness.

Each individual has an inner colour code like DNA, reflecting their karma and dharma. This is connected to their birth chart and dominant planet. It can be seen in the deeper aura arising from the spiritual heart, not simply our outer energy patterns that have emotional ups and downs. These inner colours form myriad patterns and hold certain background yantras or geometrical resonances, like the Sri Yantra. They are connected to sound vibrations (nada and mantra), and even to different fragrances.

Our outer senses only provide intimations of the wonder of our inner senses like the third eye, in which we do not merely perceive colours in outer objects, we experience colour as an effulgence of our own inner bliss or Ananda.

The power of visualising objects and internally picturing images is part of rupa tanmatra. The power of seeing subtle forms beyond the physical dimension is known as clairvoyance or clear seeing, where we can see the invisible infrastructures of forms made of colour, light and geometry and the electromagnetic field or aura around people's bodies.

When we consider that the human eye can only see between 430-770Thz, and our ears can only detect sound between 20Hz-20Khz, we can understand that these ranges make up a tiny fraction of the total sound and light frequencies in creation. This means that the vast majority of what is going on around and within us we cannot see or hear. This is why the tanmatras and their super senses are important, as with these elemental measures we can "see" and "hear" much more of the total range of sound and light frequencies, just not with our sense organs!

For example, a talented clairvoyant can see the state of the chakras in another's body, the state of the organs, the dis-ease and wellness in each of them, as well as a person's ancestors, the invisible spirits attached to people, and the colours, emotions and frequencies inside people. A talented clairvoyant can see other people's past life incarnations, and can see people as music, as light, and as colour. They can see the atom, the chakras, what blocks them, what is in them, and what illnesses people have.

Seeing people as geometry, light and colour, seeing people as music, allows us to see deeper aspects of them. We can see light in all people and all beings, freeing us from identification with their physical appearance and bodies, and we can even see others as totally empty, as formless, if we take clairvoyance to its peak.

Seers who have mastered fire element can See and Know whatever they wish to, by focusing the power of their perception onto any object, element or person, as Rishi Vasistha has showed us in his meditations on the five elements (shared throughout this book).

Today, this faculty is being explored through *quantum optics*. Physicists in Australia have simulated time travel through photons that travel along "closed time-like curves", found in the strong gravitational curvature of space-time, which bends light.

Following this curve allows one to return back to the point in time one started from, as well as before and after the point one started from! It is just a matter of time (so to speak) before refinements in optics will make time travel real by manipulating the space-time loops that forms reveal through.

Another science fiction scenario becoming a twenty-first century reality is the ability to make an object invisible by manipulating the tanmatra of form. By reflecting light through systems of cameras and mirrors, and using special materials and nanotechnology in clothes, people, cars and objects have successfully been made invisible. It is just a matter of time before buildings and bigger objects can be rendered invisible by manipulating the optics of light and sight.

Spiritually, invisibility is achieved by the yogic ability of matching one's vibration to the resonant wave reflections of the body. Another form of invisibility is cloaking one's vibration psychically, meaning people can still see your body but do not pay any attention to you as the emanations of your form are "muffled". This is a useful trick to utilize when you do not want to be noticed in a room full of people, and want to remain undisturbed and unseen, quietly going about your business.

Sounding N

This is a guttural and nasal sound. With mouth open, not vibrating the lips, with your tongue relaxed, the sound N resonates the bridge of the nose and pineal with vibratory power.

NA: Taste – Rasa Tanmatra

In the tanmatras there is a difference in their sequential order from the toroidal wave five-element sequence of throat, heart, sacral, solar plexus and root. The tanmatras follow the linear order of throat, heart, solar plexus, sacral and root, swapping the solar plexus of fire and sacral chakra of water in their manifestation.

From water element comes the tanmatra of *rasa*, which provides the potential for the experience of taste to occur.[136] It is not taste itself, as taste comes from the water element in manifestation. Water manifests the taste of whatever it comes into contact with, as it is totally permeable.[137]

A Rasa is a nectar, a juice of life, an ambrosia of delight, devotion and desire. Rasa emotionally experiences life through beauty, feelings and pleasure. Every living being, from Brahmā to a cockroach, desires to relish some sort of taste experience to validate its existence. Rasa is our taste or inclination when we like and desire something.

As Krishna in *Bhagavad Gita* states, *"Raso-hampsu"*: "I Am the taste in water", affirming that the tanmatras are expressions of the infinite. Taste is a vehicle for knowing the divine because *Rasa* is associated with *bhāva*, the first pure touch of emotion, feeling, sentiment and love we experience *without the mind interpreting it*. Curiously enough, Lord Bhava is also the presiding deity of the waters.

When we taste something, we invite it into our mouths, then into our bodies, transforming it into our own flesh, making it a part of us. This is intimacy! We taste and take things in on almost all levels of consciousness, for the sense of taste is given to us to enjoy life, and it continues the first desire of the Creator to experience and delight in Its expression in all living beings.

The next time you taste something, remember this: the taste you are experiencing is given to you so you can feel pleasure!

As Father Thomas Keating shares, "the most intimate of mystical experiences seems to be taste, in which we receive, so to speak, the kiss of God in our inmost being... the most sweet kiss pours into our inmost being, the fire, the light, the life, and unconditional infinite love, the ultimate source of creation and the universe."

Taste happens in many ways in sexual relationship, one of which is known as "the holy kiss" in Tantra, where the mouths, breath and tongues of both people lock together in an intense, holy vacuum, bringing both people into communion through the tasting of each other's flesh *and* spirit. This is the most intimate kiss possible, as both people merge and dissolve their sense of individual self through Rasa.

The tongue is the organ of taste, and is known in Tantra as the organ of desire, wishes and fulfilment. The tongue is the master of all the organs in the body and directly sprouts upwards from the heart as the embryo grows in the womb.

The tongue is a key part of tantric practice in transmuting sexual energies to expand consciousness and open the chakras in the brain through *khechari mudra*. As Swami Kriyananda shares in *The Art and Science of Raja Yoga*:

The positive and negative energies in the tongue and nasal passages (or uvula), when joined together, create a cycle of energy in the head which, instead of allowing the energy to flow outward to the body, *generates a magnetic field that draws energy upward from the body and from the base of the spine to the brain*. This energy is experienced in the mouth as slightly sweet, a pleasant *taste* that is a mixture of ghee and honey. This is known as "the nectar of the gods".

In another meditation popularised by Drunvalo Melchizedek, when a person is in a heart centred state and uses the tongue to connect to the roof of the mouth with a desire to connect with the pineal and thalamus, alpha waves fire in the four lobes of the brain, expanding one's consciousness. Many traditions agree that the tongue, taste, pleasure, love and super consciousness are interwoven together.

In the subtle sense of taste *clairgustance* or "clear tasting", one can receive taste information *without ingesting any physical source of that taste*. When clairgustants are problem solving, they can receive helpful information through extrasensory tastes in their mouths.

Sounding NA

This is a cerebral nasal sound. With mouth wide open, touch your tongue tip to the roof of the upper palate. Sound *NA* here, and the sound resonates in the bridge of the nose and pineal.

NAM: Smell – Gandha Tanmatra

Gandha becomes eternal when it is found in eternal things, and it becomes non-eternal when it is found in non-eternal things.
– Shodhganga: A study of Nyāya-vaiśeṣika categories

Earth tanmatra is the final link on the chain of manifestation, the culmination of all four previous elements and tanmatras. With the addition of the anusvara M as the sound that brings the immaterial into the material in *NAM*, all tanmatras come together.

The tanmatra of earth is *gandha*, holding the potential for the experience of smell. Gandha tanmatra is not the smell itself, but smelling is dependent upon it. Disorders in the capacity to smell are an imbalance of the earth element.

In India, offerings to the gods and devas are made through specific smells. Each god has their favourite smell, which is like food for them, as they delight in, enjoy and "eat" smells. They are delighted when their specific flower or incense is offered to them, and this delight may then be translated to the offerer who may receive the gift or boon they have been asking for.

Such non-physical beings of a high vibratory rate manifest an aspect of their frequency into the physical dimension through ethereal, ambrosial, nectar-like fragrances. These are the most beautiful aromas you will have ever drunk in, and whilst they may resemble earthly smells, they are sweeter, of a higher frequency, and are pleasurable, heavenly and intoxicating, transporting one into a supra-sensory state of pleasure.

These aromas appear suddenly out of nowhere, and this type of manifestation is a sign that a celestial being is present. If this happens to you, and it can happen anytime, anywhere, not just in meditation, stop what you are doing. Ask three times who is present. Ask them what they are here for, why they are visiting you, and thank them. Or, just enjoy the fragrant, pleasurable presence!

Non-physical beings use intoxicating smells as a bridge to manifest here, to make themselves known to us, and to bless us. Incenses are burnt as prayer offerings to connect with the gods and are what they delight in.

Experiencing these ethereal fragrances might trigger the other senses to "see", "hear" or "feel" what a spirit is trying to communicate.[138] These fragrances can also emanate from us in meditation, in deep lovemaking, in high creative states like making art or writing, and in high frequency states of consciousness.

As Father Thomas Keating said,

The first experience of God in mysticism or in contemplative prayer is analogous to perfume. You smell what you smell. If roses are there, you smell them; if God is there, you enjoy it. But if you reflect on the experience, that usually diminishes it. So you let it come, and you let it go, and don't get attached to it.

Ask yourself: What scents come to mind when you sense the divine?

In *clairalience* or clear smelling, information and insights come through the perception of smell. For example, you may suddenly get a strong whiff of the perfume your deceased mother used to wear. No one wears that fragrance in your home. Maybe your mother is reaching out from the spirit world to communicate with you. Or, maybe lately everywhere you go, you keep smelling roses, your mother's favourite flower. This may be a nudge from your guides to connect to your mother.

Most smells have emotional and mental correlations to them, and if you keep smelling the same smell inexplicably wherever you go, it is a sign from that intelligence pointing you towards an aspect of yourself you need to tend to. Smells can also trigger and evoke our memories, the strongest out of all our senses.

We receive subconscious, preverbal information through smell, with the Latin meaning of perception being "the action of taking possession and apprehension with the mind or senses". When engaged with, smell offers a direct avenue for purifying, balancing and nourishing the body's Earth Element through the limbic system.

The limbic system houses our primordial self and subconscious emotions, and includes the amygdala, which plays a key role in the processing of emotions, and the hippocampus, which is associated with memory. Recent research has also shown that we have odour receptors in the liver, heart, kidneys, skin and sperm, showing the fundamental importance of scent to our entire physiology. Our bodies are wired for scent just as they are wired for sound and touch.

Every breath we take absorbs the scents of our immediate environment, with all its resulting emotional, preverbal and instinctual cues, which we register on a subconscious level. Seeing as we smell every moment of the day, scents are a continual factor for us all the time.

Smells can be intimate, specific and personal. In the sexual act and in creative processes such as writing or painting, different pheromonal odours can exude from the body, which can be pungent or pleasant. Our scents are emotional signals, which are potently expressed through "emotional sweating" due to fear, shame, anxiety or pain, which release through the palms, forehead, armpits and the soles of the feet.

In India, there is a specific karma called odour body-making karma or *gandha karma*. When the body starts to release specific pheromones whilst in emotional states of healing, release and catharsis, this karma releases through pungent, stale, astringent and toxic smelling odours, hence the name "odour body-making karma".

Science confirms this. Utrecht University psychologists collected sweat samples from men who watched videos designed to evoke feelings of fear or disgust. Then, 36 women were asked whether they could detect any emotional cues hidden in the smells. When they were exposed to fearful sweat samples, their faces suggested fear. When they were exposed to disgust-based sweat samples, their faces mirrored disgust.

The smells of sweat are an effective means of transmitting an emotional state. Sweat is an emotional weather vane, a tool for broadcasting inner and primal feelings.[139] One aspect of smell is connected to survival and protection, for we can smell other people or animals long before we see them: we can also smell fear, tension and other primal emotions from others if we are sensitive enough to this. Our ancestors clearly took advantage of this faculty.

Smell attracts us to something or repels us. *This is a signal.* If you do not like the smell of someone, that says something! If you do not like the smell of your partner, you are not compatible on a primordial level. Smell is our most primordial subtle sense, acting as a messenger through *gandha tanmatra* to come to earth and our primal human self.

The Mirror of the Tanmatras

Sound is reflected outwardly in space element and inwardly in the ear. Touch is reflected outwardly in air and inwardly in the skin. Form is reflected outwardly in fire and in a mirror, and inwardly in the eye. Taste is reflected outwardly in water and inwardly in the tongue. Smell is reflected outwardly in the earth and inwardly in the nose.

These reflections are like the reflections in a mirror. They can only take place individually. All five reflections are not available at once, although in the case of synaesthetic people, who can smell colours and see sounds, two of these can mingle and be reflected/experienced at once.

In normal worldly life, one experiences that which is reflected in a mirror is at one place, and the mirror is at another. In a mirror, form (eyes and fire element) is reflected. Touch cannot be reflected in a mirror, nor can taste, smell or sound. A mirror will only reflect forms.

Yet, in the One Consciousness, all five can be reflected at once. Although these reflections are experienced individually in all of the sense organs, such as sight in the eye, sound in the ear etc., these reflections cannot be observed if consciousness is not there. Consciousness is the reflector, not the organs.

In the world, the object which is reflected, *bimba*, seems to be the cause of the reflection *pratibimba*, because the reflection cannot exist without that which is reflected. Yet, if the object reflected is the cause of the reflection, then what kind of cause is it? Is it an effect as well?[140]

In non-dual Saivism, all reflection is a reflection in One Consciousness. Nothing exists outside One Consciousness. There is only the mirror of One Consciousness. The universe therefore is reflected in the mirror of consciousness, not in the organs or elements, which cannot reflect anything. The reflector is consciousness.

Unlike ordinary reflection in the world, where an object is seen as the cause of a reflection, in One Consciousness only the reflection exists and not anything that is separate or reflected. In consciousness, you see only the reflected thing and not the object that is reflected. That which is reflected, *bimba*, is *svatantriya*, One Consciousness.

The universe is the reflection of this One Consciousness: the universe is found in the reflector of this One Consciousness. There is no mirror outside separate from that which is reflected in the mirror. The reflected and the reflector are inseparable. It is Shiva and his Shakti, the energy holder and his energy of will. The reflection of the universe takes place within the One.

In Kashmiri Saivism, this is explained by Shiva taking the form of a cup or grail, and then putting another cup in front of him. In the second cup, which is inseparable from him, the reflection of the universe takes place.[141]

This understanding is brought into daily use through the practice of *pratibimbavada*. Whilst walking, smelling, tasting, touching, the yogi learns to maintain an unbroken connection to the One Consciousness, and sees all One's actions move in this Consciousness. Vision, perception and supra-sensory sensation then becomes unlimited as One sees each of their actions in One Consciousness.

Sounding NAM

This is a dental sound. With mouth open and the tip of the tongue touching the lower edge of the upper teeth, *NAM* is resonated from the bridge of the nostrils and pineal.

Sounding NA MA N NA NAM

These five sounds flow one into the other and become a wave when you repeat it like a mantra. Each sound guides how the other one sounds, as they become one current of sound. Whilst they are similar sounds, the subtle differences between them all combine to be felt deeply inside you.

Namannanam is a powerful series of sounds to make as the tanmatras bring the elements into form, and one experiences these sounds primarily in the ajna pineal chakra.

NA: Tongue behind and touching two top teeth: Shabda.

MA: Mouth closes and opens with semi-pursed lips emanating *MA*: Sparsha.

N: Nasal Sound through the palate and into the third eye: Rupa.

NA: A slightly more explosive sound arising from the previous *N*, opening into *AH*.

NAM: Double Nasal Sound *N* and *M*, potently vibrating the third eye in Earth element gandha.

Tanmatra Ceremony

Ceremony is another definition of *tan*. The Five tanmatras can be unified and worked with simultaneously through:

Space and Sound: Mantra
Air and Touch: Breath
Fire and Form: Geometric Colour Meditation
Water and Taste: Devotion/emotion/heart
Earth and Smell: Mudra and Earth connection

A tanmatra ceremony starts with a person connecting to Earth Element, either by lying down or sitting with bare feet on the earth. Fragrances and incenses can be burnt and offered, and these can include specific ones for earth element, ones that have special emotional and memory significance for you, and ones that are aligned to the emotional purpose of the ceremony and what you wish to manifest.

The water element involves emotion, feeling and devotion. With eyes closed, the six Ayurvedic tastes of sweet, salty, sour, bitter, astringent and pungent can be placed on the tip of the tongue and savoured, one by one. Your tongue and taste awaken!

Desire, sincerity and deeply felt devotion to the intention and purpose of the ceremony need to be clearly felt here too through prayer.

The coloured yantra or mandala form one meditates on has to be the same frequency of the mantra you will be sounding: you will then be hearing what you are seeing and seeing what you are sounding. This brings one into a multidimensional harmony.

The breath is used in specific ways to enhance this.

The mudra employed will correlate to the mantra and geometry and connect these aspects into the electromagnetic circuits of the body.

At the end of the ceremony, one sits in silence and the purity of space element.

Now that all tanmatras are consciously felt and connected within you, one can enact this union of elements and tanmatras in ceremonial lovemaking internally, or with your partner.

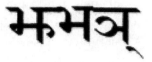

JHA BHAN:
Speech and Handling

वाक्पाणी च झभञ्ञासीद्वराड्रूपं चिदात्मनः।
सर्वजन्तुषु विज्ञेयं स्थावरादौ न विद्यते॥ १९ ॥

Vākpāṇī ca jhabhañ āsīd virāḍrūpaṁ chidātmanaḥ
sarvajantuṣu vijñēyaṁ sthāvarādau na vidyatē

19

Speech and handling evolve from the letters *Jha Bhan*,
which comes from the speech, understanding and intelligence
of the cosmic Self,
which establishes understanding and capacity for speech
in all sentient beings,
not inanimate immovable things.

Awareness presents itself in the form of all worlds,
all dimensions and in all worldly objects.

The speech or "Word" of the mighty, potent *viradrupam* Self expresses its *cit*, the fluid soul intelligence and openness through which all knowing flows. This *cit* is the establishing of the capacity within the forming foetus of direct understanding, intuition and experiential, felt and transformative Gnosis or direct cognition of anything. This is *vijneyam*.

Embryologically, this is the neurotube spark of light ignited in the embryo. On another level, the ability to speak and understand now stabilizes in our foetal form in the womb. Our speech now has the capacity to be a mirror of the Word, and we gain the capacity to create through our voice, just as the Cosmic Self does. We can now emulate through our voice the powers of creation.

Jha is speech, with its attendants in *Bhan* of coherent language, thought and neurological functions. In its true functioning in *viradrupam*, the voice can be a conduit for the four modes of sound that create the human being and the universe, ranging from the soundless sound of *Para*, the source of sound and the Word, to *vak*, the speech we make (as seen earlier in the book).

This class of sound JHA BHAN biologically attenuates us to this entire process, giving us the hardwired faculty to emulate the sounds of creation and express them.

Everything in the world of forms can be known and named through speech, as the creation stories of Brahma and Adam show. They formed the objects of the universe by naming them. In naming anything, we create it as a form. Every form has a sound, and every sound has a form.

śrutirapi "vāgēva viśvā bhuvanāni jajnē" iti
sūkṣhmā vāgēva viśvākārēṇa pariṇamatē vivartatē vētyarthaḥ
– Tattvavimarśinī 05-06:b

Śruti (Vēdās) says, "It is speech which created the worlds." It is the subtle aspect of speech that transforms itself into the forms of the worlds.

Śrutyantaramapi "vācaiva viṣvaṁ bahurūpaṁ nibaddhaṁ tadētadēkaṁ pravibhajyōpabhunktē iti"
– Tattvavimarśinī 05-06:c

"It is said that the world with its multiple forms is composed of speech, and having divided, one part of it is being experienced."

The science of speech reveals through seed syllables or *bijas*, the foundational units of speech, the seeds from where all speech arises. As individual seed syllables unfold and join with each other, they create complex forms through the combining of light bearing vowels and form making consonants. Consonants insert between vowels, and the resulting sonic signature of these harmonic clusters become forms.

JHA and *Vak* is the process of how sound creates the forms of the universe through the voice or Word of the Creator, which has now established this consciousness in us via our capacity to speak and understand. Our speech can be an echo of the Word, an echo in human form of the One Source expressing itself. Speech can be an instrument for transmitting pure consciousness.

Our voice is the most powerful tool we have to create, the first instrument that gave rise to all music and all other instruments. The more we own our voices and words as creators, the more we can shape our lives in harmony with the original Word of creation.

When your speech becomes this medium for the Cosmic Self *viradrupam*, it deeply impacts yourself and others. You can influence people and events through the vibrations of your voice, dramatically so when your voice is aligned to *citmanah*, the soul intelligence, and *vijneyam*, felt, direct knowing and intuition.[142]

Sounding Jha

Jha is a palatal sound, pronounced similar to "jar" or "jah" with emphasis on the JH. Your tongue touches the top of the lower teeth with JH. Your mouth opens as you put more air into the *AH* sound after *JH*.

Bhan: Handling, Grasping, Understanding

There are two struggles: an inner-world struggle and an outer-world struggle... you must make an intentional contact between these two worlds; then you can crystallize data for the third world, the world of the soul.
– Gurdjieff

Vak is the seed sound for the organ of action speech. Speech needs subtle thought processes in order to make sounds, and *Bhan* is the neurological capacity to formulate and string together the units and sounds of speech.

Bhan is the function that handles and grasps thoughts and ideas, formulating them, sequencing them and giving them coherence, to then manifest them into the world through speech by pairing motor actions to these neurological functions.

Bhan is the seed syllable sound for how we grasp and understand thoughts and ideas, to then translate these into words and speech expressing our inner perceptions. The word "perceive" comes from the Latin *percipere* – to grasp.

Our grasping of stimuli and information, which we then translate into perceptions, helps us to find our orientation, our inner equilibrium, and where we stand (*sthau*). It is part of our internal mental safety mechanism. This becomes more interesting when we understand that *Bhan* is the Air element's organ of action, connecting to touch, understanding and handling.

This is *Pani*, the principle where we hold, "catch" and comprehend ideas and thoughts. Pani acts as a bridge between our internal and external worlds, between what we perceive is inside us and what we perceive is outside us. Through handling we perceive differences and learn how to operate within these boundaries. We differentiate between what is inside oneself and what is outside oneself, and how to express this. For example, we know when we are speaking aloud and when we are thinking internally. To mix up the two can be awkward!

Pani is handling. Through our hands we learn how to operate and express our inner understandings and volitions. "We are only able to fulfil our purpose if we carry our principles into everyday life. We must put them to use in such a way that every turn of the hand, every movement of a finger, is an expression of the spirit."

Your spirit and character are defined by what you do with your hands, will and actions. Who we are is partly defined by how we express our inner life in the outer world through our handling, understanding and speech.

The hands are organs of movement and organs for the spiritual. Our hands express the power of action to create from pure consciousness, and our hands literally express and shape consciousness. In our natural blueprint of connection to the quantum code, they can express quantum consciousness.

The hands are sometimes compared to the eyes: they see. In modern psychoanalysis, when hands prominently appear in dreams it is acknowledged to be the equivalent to the eyes, serving the psyche to *relate the contents of the unconscious to the conscious mind, the inner to the outer world*, acting as a bridge of connection.

Our hands provide major keys to what is happening in our life, from reading the palms, which is a key part of Vedic Astrology, to making mudras, which change the flow of electromagnetic energies within your meridians and *nadis*.

Energy flows through our palm chakras, and the hands *BHAN* are connected to the vishuddha throat chakra *JHA*, the chakra of sound and speech. The vishuddha chakra is an interface between what lies inside us and what we express externally. The throat chakra and the hands connect in a circuit.

Our hands are powerful interfaces linking our internal world to the outside world. The movements of the hands form letters in many traditions, such as the Hebrew, Runic and Sanskrit (in the Sri Chakra tradition) all of which have hand mudras for each letter.

The Hebrew alphabet has been proved by scientist Stan Tenen to emerge from the movements of the hand. He says, *"the Hebrew alphabet links the inner world of the mind with the outer world of experience, just as our hands do."*

This function of being able to distinguish inside from outside is what our hands express and help us distinguish, navigating the boundaries of where we end and where "others" begin. This distinguishing of inside from outside is the most basic distinction you can make at any level of consciousness.

The hand contains a map of all the body's energy systems in one part, like a hologram. All meridians, seven chakras, five elements, the polarity of male and female, and the measuring of time itself as seen in the Mayan Calendar system use the hands in order to *measure and map consciousness.*

"The basic pattern forming processes of nature, which have shaped the human hand and mind, can continue to guide whatever the hand and mind are shaping when the hand and mind are true to nature. The Greeks thought that humans had the capacity to reflect limitless harmony and beauty, conceived as divine. Thus, man was said to 'be the measure of all things'," as 5th century BC Greek philosopher Protagoras proclaimed.

For thousands of years, humans have measured creation through our hands in order to build and create. For instance, in ancient times the length of a man's forearm with the hand outstretched was called a cubit. The Egyptians had a smaller cubit which consisted of 6 handbreadths, and a larger one called the Royal Cubit which was 7 handbreadths. The Egyptian "hand" was made of four fingers or "digits", and a further measure, the "fist" was 1.333 handbreadths, used to establish the proportions of their royal statues.

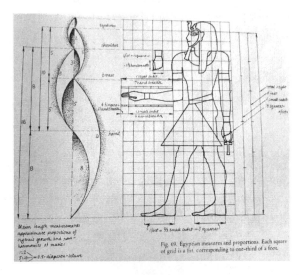

Greek Metrological Relief. Distance from fingertip to fingertip is a fathom; footprint above figure indicates basic measure.

The mapping and measuring of creation comes from understanding the proportions of the hand. This measuring of length helps us to build, so we could say that our creation comes from our hands! So too comes the measuring of numbers from our 10 fingers or *digits*, which enables us to understand the cosmos and how it works.

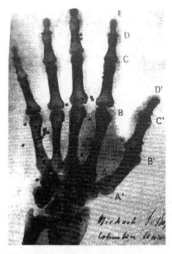

From **The Power of Limits** *by G. Doczi.*

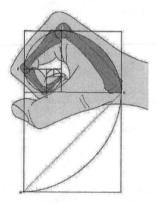

The depth of sacred geometry is literally in our hands. The phalanges of the human hand are aligned to the divine proportion or Fibonacci Sequence of 3 is to 5 is to 8 is to 13 and onwards. The living mathematics of infinity are coded in our knuckles. The lines of the hand match the geometry of nature and leaves. Our fist makes the golden spiral naturally. Jain 108.

The Science of Speech and Handling

Bhan or handling in Sanskrit has a deeper significance because Sanskrit sounds are *vibrations that points to qualities* more than objects. Objects arise from the wave-like qualities of vibrations. To translate these wave forces into language requires a control panel and interface: *BHAN.* We grasp and comprehend on a neurological level before we can formulate words and thoughts.

Science has recently shed some light on this. In the January 2020 issue of *Nature,* neuroscientist Daniel Margoliash at the University of Chicago shows how the brain is organized through sound, leading to new ways of understanding speech. His work *"provides new insight into how the physics of producing vocal signals are represented in the brain to control vocalizations,"* says Howard Nusbaum, professor of psychology and an expert on speech.

They discovered that as a bird sings, its neurons prepare to make the next sounds while other neurons synchronize with the current notes being sung. Human singers do this too, and in this coordination of physical actions and brain activity, we find the ingredients needed to produce complex movements.

By decoding the neural (Bhan) representation of speech (Jha), we can understand how the brain and body carry out complex movements, from throwing and catching a ball to doing somersaults. By looking at the physiological variables a bird uses to control singing, *"one fascinating observation surprised us: the forebrain neurons fire precisely at the time a sound transition is being produced. The timing of this suggests they are evaluating feedback from the sound."*

Whilst this is unsurprising to a human singer, science is now proving that this level of understanding is essential in coordinating speech, and there is now a mathematical description that matches brain activity for highly skilled behaviours.

Another study has shown that we recognise the intended meanings of basic sounds made by others to represent specific objects and actions – regardless of the language they speak. These vocalizations, such as the imitation of snoring to symbolise sleep, or roaring to denote a tiger, could have played a crucial role in the development of the first human languages.

This finding contrasts with another assumption: that physical gestures and signals drove the development of human language. "People around the world, whatever their linguistic or cultural background, were remarkably good at being able to guess the meanings of these different vocalizations," said senior author Marcus Perlman, a linguist at the University of Birmingham. "This could have big implications for how spoken languages got off the ground."

Until now, researchers had assumed that human languages developed through the use of iconic gestures – such as wiggling your arm to mimic the movement of a snake – and other physical signals, Perlman said. After communicating with gestures, early humans would then have gradually added spoken words that would have replaced these physical signals, according to this theory.

"It makes sense," Perlman said. *"If you go to a country where you don't speak the language, the intuitive way to communicate is to gesture what you're trying to express."* However, our ability to interpret the meaning of iconic vocalizations suggests humans may not have needed physical gestures to create words. Instead, vocalizations may have been the first building blocks of languages, and physical gestures may have been added to individual words afterwards, Perlman said.

"A more compelling argument for the role of iconic representation in language evolution comes from manual gestures," said Michael Corballis, a psychologist specializing in language evolution at the University of Auckland. *"Sign languages have a more obvious*

iconic element than speech does. Although, there is increasing evidence of an iconic component in human speech," Corballis said.

"In reality, the development of the first languages would have taken thousands of years, and it's likely that a combination of vocalizations and gestures played a part," Perlman said and continues: *"We have hands and a voice, and we have been communicating with both for many millions of years."*

"I agree that a multimodal origin is the most plausible," says Michael Arbib, a language expert and computational neuroscientist at the University of Southern Carolina. *"Certain entities have distinctive sounds which favor the use of sound symbolism for their origin, whereas many others are more hospitable to pantomime."*

In another understanding, our first and primary method of communication was telepathy, followed by using our hands and voices. As one of the main functions of language is to convey meaning, researchers from the University of Western Australia tested a broad range of people to see whether gestures or non-verbal sounds were more effective at getting meaning across.

Successful communication was *twice as high* through gesturing rather than vocalizing, as University of Western Australia cognitive scientist Nicolas Fay *explained on Twitter.* *"These findings are consistent with a gesture-first theory of language origin, and gesture is more successful than vocalization because gestured signals are more universal than vocal signals."* So, before we came up with words, more complex meanings may have been articulated better by our *hands* rather than our vocal cords.[143]

But as with the chicken and the egg, it is hard to say which came first: vocalizations or gestures. Both are intertwined and help form and inform each other, as Jha Bhan shares. They come together as a package, and are not separate. The twinned function of *Jha Bhan*, speech and handling, coexists on a fundamental level.

To try and fully understand speech Jha without the component of handling Bhan is like having your right hand working and the left hand tied behind your back. They intertwine on the most formative of levels from our time as a foetus in the womb, enabling us to express intelligence, understand, have intuition and "gut" feeling.

Sounds and letters, language and thoughts, are intimately intertwined with the modelling-building function and interface of our hands. Understanding both speech and handling, Jha and Bhan, is crucial to the origins of human language and the future of human language.

This is important because language is a huge key to what it means to be human. As Perlman concludes, *"Language speaks to the human condition, our history, our relationship with the world around us, and the essence of who we are."*[144]

Mirror Neurons

How many times have we imitated our heroes and heroines, because we want to be like them? How many times do we copy the mannerisms, language and lifestyles of those we admire and want to be like? How much are we like our parents, our first role models we learned from and imitated? How much do millions of people imitate the influencers on social media they want to be like?

The basis for this imitation learning function is found in our mirror neuron system, which "mirrors" the behaviour of the one we are observing, *as though you yourself (the observer) is acting*. Mirror neurons enable us to grasp and imitate the behavioural movements of others, creating a neural map in your brain of another's intention: you see the intention behind the behaviour, predict it and replicate it.

"Mirror neurons suggest that we pretend to be in another person's mental shoes," says Marco Iacoboni, a neuroscientist at UCLA School of Medicine. "In fact, with mirror neurons we do not have to pretend, we practically are in another person's mind." As Dr Dan Siegel continues, "mirror neurons show that we are hardwired to perceive the mind of another being."

Since the discovery of mirror neurons by Italian scientist Rizzolatti, who recorded cells in the ventral pre-motor area of monkeys' brains firing when they were using their hands to grasp, pull, push and handle objects, mirror neurons have been implicated in a broad range of phenomena, providing a basis for mind reading, empathy and imitation learning.

Anytime you watch someone else doing something (or even starting to do something), the corresponding mirror neurons can fire in your brain, allowing you to "read", understand, and respond to another in an exponential feedback loop so that you become more like them.

Once you have these two abilities in place (which we all do) of the ability to read someone's intentions, and the ability to mime their vocalizations, then you set in motion a rapid increase in your ability to adapt and learn. For example, if you want to learn someone else's movements or speech, you focus on them and relate to the corresponding movements in your own brain at the same time; without mirror neurons, you cannot do this.

Sound provides a key to this process. For example, if you closely watch someone chanting, through mirror neurons you can silently chant what he or she is chanting at the same time without even knowing or understanding what they are saying. This allows you to literally become the chanter and the listener simultaneously!

This may sound incredible, but it works: you become that sound vibration and access that vibratory wave directly. You are passive and active simultaneously, One with the whole process. This is one of the secrets to successful sacred evocation and ceremony.

The first time I did this was totally spontaneous. I was at a fire ceremony or *yajna* in India, where priests were very rapidly chanting mantras into a fire-pit. I watched their mouths and movements closely, and without intellectually knowing what they were saying *at all*, I began to silently imitate what they were chanting.

As I entrained with their sound vibration, my state of consciousness shifted from being a passive listener to an active participant, which shifted my consciousness dramatically. In effect, I started to lead the ceremony *with them*, and receive the benefits from it *much more* than if I had been passively watching them, as everyone else was doing.

Another example of mirror neurons is when I was invited to Dr John Reid's lab in England for some sound experiments in sacred languages with his invention, the CymaScope. This was in the initial stages of the development of this technology, which translates sound into visual shapes and geometries by recording how sound vibrates water into patterns.

The CymaScope had a camera projecting the geometries made by the sounds I was singing in Sanskrit, Tibetan and Aramaic onto a wall. As I gazed at the geometric shapes being produced, these shapes began to feed back to me their resonant information. This in turn informed my brain, and I automatically began to respond to this feedback, changing my chanting until the sounds I was making were quite different to the ones I started out making.

The visual geometries produced by my singing actually began to inform me of how to sing, changing the way I shaped my lips and vocal chords to produce the sounds. My consciousness altered in this feedback loop, as I implemented how the information of the shapes were directing me to sing. Previously I had intellectually understood that sound was geometry, but this was the direct experience!

We can experience this in everyday situations with others by watching them, their movements and their words, listening deeply. In a learning environment, we can listen to a teacher's words and intentions closely, and set our intention to trigger our mirror neurons *to become the information being transmitted*. This can then become an *interactive* learning experience.

Mirror neurons activate when you perform an action, and when you observe the same action performed by yourself or another. Have you ever stood in front of a mirror practising speeches, conversations or facial gestures, or even saying "I love you" to yourself? This is activating your mirror neurons as well.[145]

Mirror neurons enhance social connection and empathy with others. We can grasp feelings and see things from another's perspective. We are natural mind readers, and we can place ourselves in another person's "mental shoes", using our own mind as a model for theirs.

Neuroscientist Gallese contends that when we interact, we do more than just observe the other person's behaviour; we create internal representations of their actions, sensations and emotions within ourselves, as if we are the ones that are moving, sensing and feeling.

We share with others the way they act and subjectively experience emotions and sensations, and we also share the neural circuits enabling those same actions, emotions and sensations: the mirror neuron system. In other words, "you either simulate with mirror neurons, or the states of others are completely precluded to you," said Iacoboni.

Mirror neurons could explain big changes in human evolution to a degree: inventions like tool use, art, mathematics, and aspects of language may have been invented in one place and then spread quickly given the human brain's amazing capacity for imitation learning, transmission and mind reading using mirror neurons.[146]

Mirror neurons also relate to the evolution and exponential growth of interpersonal communication and social intelligence. Using the mirror neuron system results in us being able to quickly learn, absorb and transmit information we could not do alone or by ourself. Mirror neurons open a road to understanding the bridges between minds, and how we are influencing each other's development through relationship, social media and community resonance.

Awareness Is Inside and Outside

When you grasp that what is within you becomes projected outside you, then you can cognise yourself. When you can handle that what lies within you that is unexpressed or repressed can become projected outside you, then you can see more of yourself. When you understand in your open, feeling soul intelligence that you project outside of you what is within you, then you can cognise yourself.

This leads to Gnosis, the direct, experiential, felt and transformative understanding that is *vijneyam*. This ultimately leads to one cognising that there is a single awareness intelligence operating within you, outside you, through you, and in everything.

As Rupert Spira shares, *"There is no separate inside self and no separate outside object, other, or world. Rather there is one seamless intimate totality, always changing when viewed from the perspective of objects, never changing when viewed from the perspective of the totality."*

Sanskrit words do not take the perspective of objects: they are vibrations of qualities and characteristics, which may include objects but which in and of themselves are not objects. When our speech *Vak* becomes a medium for this awareness in the unbroken continuum of Self-expression, we come into harmony with the cosmic self.

Jha Bhan Sound Practice

BHAN is a labial or lip focused sound. To sound *BHAN*, your tongue is relaxed and your lips are rounded and slightly open. Sound *BH* from this position, and with the *HA* sound, add more breath. The *N* nasal sound vibrates the pineal. Remember to keep the lips open so the lips do not vibrate, but rather the sound vibrates the palate and pineal.

1. Sound *JHA* behind the right ear x22, then internally x22.
   Sound *BHAN* behind the left ear x22, then internally x22.
   Now: Sound *JHA* in the right forebrain x22, then internally x22.
   Sound *BHAN* in the left forebrain x22, then internally x22.
   Now, repeat *JHA BHAN* at the same time, focused on both right and left forebrains simultaneously. Do 22 times aloud, and 22 times internally.
   Sit in this vibration silently for a few minutes. What do you feel?
2. Sound *JHA* in the throat chakra x108.
   Sound *BHAN* into both palm chakras x108.
   Sit in the vibration silently for a few minutes. What do you feel?

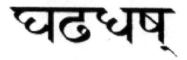

GHA DHA DHASH:
Movement, Excretion,
Procreation

वर्गाणां तुर्यवर्णा ये कर्मेन्द्रयि गणा हि ति।
घढधष् सर्वभूतानां पादपायू उपस्थकः॥ २०॥

ghaḍhadhaṣ sarvabhūtānāṁ pādapāyū upasthakaḥ
karmendriyagana hyete jata hi paramatmanah
vargāṇāṁ turyavarṇā yē karmēndriyam aya hi tē
20

Gha-dha-dhas exists in all the elements
as the forces of movement, excretion and procreation.

This completes the group of the organs of action,
the active powers birthed from the Supreme.

The syllables that are the fourth letters in their groups
manifest as the motor organs of action.

(gha, dha, ḍha, ṣha are the fourth letters of
ka, ṭa, ta and sa groups in the Sanskrit alphabet)

Gha Dha Dhash marks the end of the first part of the Sutra in the ninth beat of Shiva's Drum. If you remember from the first verse of the Sutra, Rishi Nandikeshvara shared that *"After his Dance, Nataraja Raja, Lord of the Dance... beat his drum 9 and 5 times."*

The nine and five sounds from Shiva's *Damaru* Drum emanate forth the sonic blueprint for creation and the human being, resounding these sound vibrations into being. These sounds spring forth and weave this web of vibration that arises from Nada-Bindu,[147] the Singularity point and origin of creation.

The ninth beat is Gha Dha Dhash, which *"completes the active powers generated from the Supreme Reality"*, implying that these first nine classes of sound are direct emanations of consciousness marking out the causal phase of creation.

The first three classes of sound form the quantum field, which then blooms through the fourth class of sound into the big bang *AIAUch*. The next two classes of sound unfold this torus wave spiral, forming the five elemental powers that bloom the quantum field into physical creation through hydrogen, oxygen and nitrogen. This culminates in the forming of earth element and carbon in *LAn*, the sixth class of sound.

The seventh class of sound is the number of completion of a cycle. Here the elements explicate themselves into form through the tanmatras, the measures and threads of the elements, which expand outwards into creation as shown on the diagram below.

The eighth and ninth classes of sound show *the embodying of the pure sentience of the quantum field into physical form* in the developing foetus. This occurs through the movements of the toroidal wave *flowing into* and forming the mouth and the capacities for speech and understanding in the embryo in JHA BHAN, and now, in the ninth class of sound GHA DHA DHASH, the movements of the toroidal wave *flow out* of the foetus, forming the anus and sexual organs.

In the movement of GHA our body-minds connect to this toroidal wave, which then moves to form the energetic function and anal organ of excretion/release DHA, and then forms the energetic function and sexual organs of procreation in DHASH.

In this in-and-out flow process, the toroidal wave forms within us the power and capacity to reproduce and multiply ourselves. This culmination of the 9 Classes of Sound (the numbers above each diagram) and the completing of the nine steps of the original transcription and sequencing code of the One *"completes the active powers generated from the Supreme Reality"*. This completion from the quantum field into human form in an unbroken continuum happens via the Fibonacci sequence. (The numbers below each diagram.)

From here on in our embryological evolution, we can now duplicate and replicate ourselves in this continuum, but with the added code of sexual-physical cell division and mitosis included in this evolution.

This entire causal creation sequence is remarkably similar to the way the Flower of Life Code operates. The Flower of Life is a geometric sequence found in over 60 countries worldwide, including India, and shows how life unfolds in geometric progressions from the Creator or Void through the Phi Spiral, Golden Ratio and Fibonacci Sequence.

The quantum code of the Maheshwara Sutra is the progression of the "Fibonacci" sequence, which adds itself to itself in order to evolve its code. This spiral sequence continues infinitely and has no end, as mathematicians have discovered.

The "Fibonacci" sequence was discovered by the Vedic grammarian and linguist Sage Pingala (i.e. the numbers were discovered through sounds) 1600 years earlier than Fibonacci, in the 4th century BC.[148] This sequence became the basis of Indian classical music and Sutra composition in his *Chandas Sutra*, which has been used for thousands of years in the composition of Sanskrit prosody, verse, poetry and mathematics.

In 1050 AD, the scholar Hemachandra rigorously codified this sequence. Thus, the Fibonacci sequence should rightly be called the *Pingala-Hemachandra sequence*.

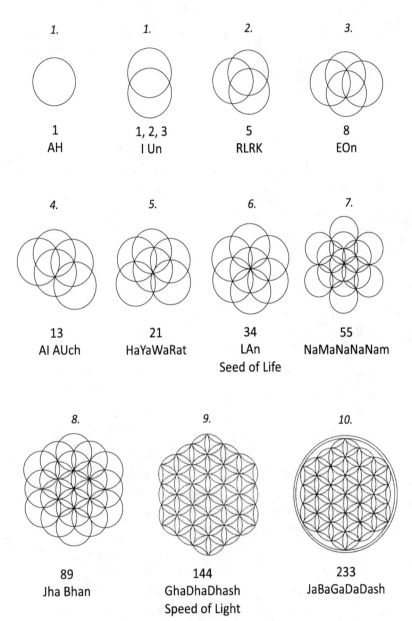

1.	*1.*	*2.*	*3.*
1	1, 2, 3	5	8
AH	I Un	RLRK	EOn

4.	*5.*	*6.*	*7.*
13	21	34	55
AI AUch	HaYaWaRat	LAn	NaMaNaNaNam
		Seed of Life	

8.	*9.*	*10.*
89	144	233
Jha Bhan	GhaDhaDhash	JaBaGaDaDash
	Speed of Light	

In the diagram above, we see that the Flower of Life geometric unfolding of creation is identical to the sounds and unfolding of creation in the Maheshwara Sutra. Both unite through the universal languages of number, geometry, sound and meaning.

Each diagram shows the geometric unfolding of creation from *AH*. The numbers at the top of each diagram are the 9 steps of creation: the numbers below are the numerical progressions of the *Pingala-Hemachandra sequence.*

The Flower of Life and the Maheshwara Sutra

1. A I U n = 1, 1, 2, 3

The 1 Appears; then the 1 becomes The 1. 1 + The 1 is The 2, which marries itself in indivisible quantum unity to form the "holy" trinity 3 that begets all of creation. 2+1=3. These are 4 movements.

A I Un is a unity of 3 vowels + 1 consonant *n*, which manifests quantum matter. (Consonants create form, vowels are light.) It is interesting to note that the 4 sound-number movements have 3 circles in 2 diagrams in the 1 sequence of *AIUn*.

2. RLRK = 5

3+2=5. Each class of sound apart from the first class *AIUn* follows a more regular pattern of having a single number for itself.

3. EOn = 8

5+3=8, the number of original encoding cells in the foetus, the Egg of Life geometry, and the formation of the primordial egg from where physical creation arises. If these are 8 beats, then in the Hemachandra-Pingala sequence we can have 34 combinations of long 2 beat sounds and short 1 beat sounds at this stage of creation, i.e. 34 is the total number of possible rhythmic combinations. 34 is Earth element, the manifestation into form of this sequence later on in the Sutra.

4. AIAUch = 13

8+5=13, the magic number of physical creation appearing and manifesting from the invisible quantum realm in the Big Bang.

5. HAYAWARAT = 21

13+8=21. The four elements of space, air, water, fire manifest.

6. LAN = 34

21+13=34. Earth Element manifests. Between the harmonics of 21 and 34, as seen in a sunflower's counter-rotating spirals of petals, we find the Golden Ratio of perfect harmony. So, between the creation of the four elements and earth element the Golden Ratio is seen clearly as the perfect harmony of the elements.

7. NAMANNANAM = 55

34+21=55. This is the explication of the 5 elements into more complex forms in the universal unfolding of matter, *and* the foundations of the subtle or super senses in humans. This unfolding and explication is clearly seen in the diagram.

8. JHA BHAN = 89

55+34=89. The forming of the mouth, speech and understanding functions in the neurotube of the developing foetus are fundamental markers of sentience.

9. GHA DHA DHASH = 144

89+55=144. The magic number of 144 is where our ability to replicate ourselves through sexuality manifests. The first 9 beats of Shiva's Drum now complete the causal phase of creation, and the Flower of Life completes its causal phase *as well*.

144 is a magic number in many sacred traditions, as it is based on the living mathematics of the quantum code, which is reflected in nature. The 9th Class of Sound in the Maheshwara Sutra and the 12th number of the Hemachandra-Pingala sequence is 144, which is *the only number in the sequence which is a square of itself.*

This leads into another mathematical harmony. Between 89 and 144 lies the Golden Ratio of 1.618. Thus, if you divide 144 by 1.618 = 89, divide 89 by 1.618 = 55, and so on. The Hemachandra-Pingala sequence cascades into the Golden Ratio at 144.

The DNA molecule is characterised by the geometric shapes of a pentagon 5 joining a hexagon in 6. If you square these numbers into 36 and 25, and then divide 36 by 25 you get 1.44. In the pentagonal geometry of our cell proteins (which promote growth), which fractally recur in the Golden Mean in our atoms and in the galaxy, 144 is encoded.

If you count all the petals within the *entire flower of life* (each individual unit of the flower of life is a hexagon of 6), and the negative spaces between the petals, there are 90 petals + 54 spaces = 144.

In the Vedic mathematical phenomenon of magic cubes, specifically the 6x6 cube, all the numbers add up to 651. There are 144 ways in which this magic cube can be formed. The cube form stores information. In the Bible the measurements for the new Jerusalem or heaven, the temple for humanity, are based on the perfect proportions of the human being.

Guess what? Its measurements are based on 144 cubits. This perfectly proportioned temple of heaven and the perfect human being in harmony was also envisaged to surround the earth and its grids, meaning the human being in symbiotic harmonious relationship to the earth.

1/144=0.0069444. The brilliant inventor Nikola Tesla built antennae of 6.94 feet based on this harmonic of light as a basis for free and unlimited energy.

In dividing a circle into 10 equal parts with 10 angles, we get a decagon or ten-pointed star with angles of 144 degrees each. If you pull out a 1/10th part of this decagon, like a pizza slice, you see a Pythagorean golden triangle. 144 is made up of the two 72 degree sides of this golden triangle, in the golden ratio of 1.618.

The 144-degree angle *brings the golden ratio into the shape.* Building anything with a 10-sided foundation in this way encodes the golden ratio into the structure. Nobel winning physicist Roger Penrose also discovered that the 144-degree angle is the *key to atomic structure* in his "Penrose tiles", also known as "aperiodic" tiling.

In Vedic times, 1 kala or unit of time was 144 seconds. They understood that *timing measurements of 144 units are in harmony with universal constants*. In Vedic spirituality, the crown chakra of an enlightened person has 144,000 petals, meaning they were "in light", hence the modern term "enlightened".

Perhaps the most important aspect of 144 is that it is the *frequency for the speed of light*. In the earth's grids, 144 is the number of minutes per arc grid second, meaning 144 is the speed of light. Bruce Cathie did extensive mathematical measurements to prove this.

This speed of light 144 is the speed of light in *the relative or physical universe*. In Einstein's theory of relativity, it is not possible for any object with "ordinary mass" to move faster than the speed of light. The speed of light operates as a speed limit on anything with "ordinary mass". Ordinary mass is simply how we weigh objects using current scientific instruments.

To accelerate an object with ordinary mass, we have to add energy. The faster we want the object to go, the more energy we need. The equations of relativity tell us that anything with ordinary mass requires an *infinite* amount of energy to be accelerated to the speed of light. This of course is all based on relative laws and measurements of the physical universe, not quantum laws or unified theories.

However, certain particles called *tachyons* can travel beyond the speed of light because they have "imaginary mass", and are *always* travelling faster than the speed of light. Just as something with ordinary mass cannot be accelerated past the speed of light, *tachyons cannot be slowed down to below the speed of light.*

Some physicists believe tachyons constantly travel backwards in time.[149] If this is true, tachyons could travel back to the beginning of creation, to the beginning of the quantum code *AIUn*, acting as a link of light to the quantum code.

This is why the demarcation between the 9 and 5 beats of creation is so relevant: they mark the beginning of relativity, mass and physical laws as opposed to quantum laws which have no mass and obey non-physical laws through quantum particles like tachyons.

Thus, GHA DHA DHASH could act as a wormhole out of the relative reality of the physical universe through this boundary point. A wormhole is a gateway or shortcut between any two points in space.

Whilst a star might be 9 light years away in relative terms, it might only be a 5 hours away via a wormhole. Wormholes allow travel over great distances in very short periods of time – allowing us to get to the farthest reaches of the Universe within hours.

The relative speed of light that is formed in GHADHADHASH links to the quantum code that created it, creating a wormhole or bridge from the quantum code of the first 9 massless quantum classes of sound and 9 steps of creation, to the relative physical universe.

Beyond this relative speed of light 144 lies the powers of the 8 previous classes of sound: the quantum code that is non-local, beyond relative measurements of mortal life and death, is instantaneous, "beyond" and "before" relative speed, time and distance. It cannot be measured with the senses or mind.

This tachyon bridge into the quantum universe is a two-way mirror, depending on what state of consciousness you observe or access it from. It is a portal into the quantum code of the One in the first 9 classes of sound, *and* it also reflects and forms the finite physical universe on the "other" side of the mirror (as explained in the next 5 classes of sound), completing the quantum code into physical embodiment.

Gha Dha Dhash

In GHA DHA DHASH is where *Paramatmanah*, the One, forms the *karmendriyas* or organs of action in our bodies: the anus, the reproductive organs of lingam/yoni, and the power of movement in the feet and legs.

GHADHADHASH is the *movement between, in and through the legs, lingam/yoni and anus. These are three organs but also three actions which exist in every cell,* as they move and vibrate, release energy and reproduce energy.

GHA is the seed sound of the *Karmendriya* organ of action *Pada*: movement and locomotion, sometimes referred to as the feet and legs but also the very power of locomotion itself. GHA is movement arising from the fire element (the tanmatra of form *rupa* as we saw in the 7th class of sound) creating form through motion. This GHA motion continues the toroidal wave of AIAIUch, but is now powered by the fire element.

GHA is a strong movement, a winding spiral pushing inwards with pressure, thrusting and spiralling. Its semi-guttural sound when pronounced, indicates this strength.[150] *GHA* continues the previous class of sound *JHA BHAN* through another twisting of the toroidal wave movement (that previously formed the mouth JHA) to now form the anus, in the outflowing toroidal wave *DHA*.

DHA has a softer movement that continues the action of *GHA*. *DHA* means to place, fix upon, direct towards, generate and *release/excrete*. So, the movement force of GHA going inwards brings forth DHA in order for the final motion into *DHASH*, the sexual, generating urge to reproduce through the sexual organs. This is all one interconnected toroidal flow: movement-excretion-procreation, forming the anus and lingam/yoni.

DHA arises from the water element and releases energy, allowing the full unrestricted flow and cycling of the torus spiral out of the form it has just created. Water lets go of all that is stuck, stagnant and blocked into fluid forms that are adaptable.

Grasping nothing, water flows and does not hold onto anything for longer than it is needed. Water is fully pervaded by whatever it comes into contact with, and *Payu* hints at water's ability to always be in its natural flow, ever free. It permeates everything, and when it leaves anything this giving up of association is *Payu*.[151]

As this energetic function develops later on in the growth of the foetus, it also lets go of energies and wastes which are no longer useful on all levels physically, emotionally and mentally, through the pranic flow known as *Apana*.

Letting go and excreting "comes out" in many forms. From the release of fluids and sweat to cool our bodies and release toxins, to the crying of tears to release emotions, to releasing mucus to free our lungs, to the release of faeces and urine to keep our bodies cycling and functioning properly, we need to flow and release to remain alive and to process wastes we generate every day.

Excretion, letting go and release are part of the foundation of our physical, emotional, mental and spiritual well-being, survival and growth. We need to release physical wastes, otherwise our bodies will block up and eventually die. We need to release our emotions otherwise we become emotionally repressed and unable to enjoy love and life.

We need to empty our minds and our mental inboxes in order to relax and grow, otherwise we become rigid, fixated, closed and fundamentalist. We need to release our spirits through meditation and prayer otherwise we cannot expand into the infinite. *DHA* is a foundational building block for our body-minds and souls. What goes in must come out, just as energy is never destroyed, just transformed into different states.

DHASh or Procreation is the final movement in *Gha Dha Dhash*. It is the seed sound of the organ of action *Upastha* (meaning upstanding) or lingam, which generates all procreative activity and is the biological and sexual urge to reproduce, *to be for* (pro) creation.

DHA means upholding, supporting and nourishing creation, and any activity that promotes, reproduces, generates and creates more of itself. Procreation is forward movement designed to reproduce more of itself. The *"organ of procreation is the one goal of all kinds of enjoyment."*[152]

The procreative energy now fully materialises with DHASH. The toroidal motion of GHADHADHASH now completes by connecting to earth element as the final stage unfolding from fire GHA and water DHA. Earth is the fuel and food for all the elements, the food for the individual self and the fuel food for semen as the substance of procreation.

From the seed of semen comes *the seven dhatus* or foundational building blocks of the body which give rise to *Ojas*, our strength, vigour, vitality and virility. The more Ojas we have the more powerful we are, having charisma, impact and power in what we say and do.

Ojas is sexual energy that can be refined into subtler and more potent forces within us in order to accelerate our spiritual development. When our sexual energy *and* sexual thoughts are refined we gain more Ojas, which is guided by the muladhara root chakra and earth element.

This sublimation "positively, dynamically converts and controls the sex energy, conserving it, diverting it into higher channels, and converting it into *Ojas Shakti*. The material energy is changed into spiritual energy, just as heat is changed into light and electricity."[153]

We can transmute sexual energy and thoughts into a clear, grounded, self-contained and elevated consciousness. This allows our creative expressions to flourish and deepen as the pro-creative power is channelled into other parts of us, such as fulfilling your purpose, your dharmas, and accelerating your own growth amongst many possibilities.

By transmuting sexual energy, we have more energy to fulfil other aspects of us, whereas before you may not have had enough energy to do so.

Gha Dha Dhash are three interconnected actions that move, direct, release and ground the procreative energy of the toroidal wave in the foetus. The fire, water and earth elements cycle in each other at different vibratory rates within the toroidal spiral, activating the primordial sexual code of reproduction.

On another level, GHA is a movement that culminates in a mini death in DHA excretion, and then re-creates life again in the procreative power of DHASH. Here, the truth of mortality is expressed: what is born must die. The death that the power of mortal sexual reproduction brings is inevitable, yet it also brings us the capacity to be creators.

Yet, sexuality is a Tantric pathway into the quantum code of the One *if done consciously*, AND in this 9th step of GHADHADHASH mass, relative physical laws and mortality start to be introduced into human evolution, which previously did not exist in the quantum code.

This is why the Maheshwara Sutra is divided into two parts of 9 and 5. The first 9 movements of the quantum code are untouched by mass, time, death, relative physical laws and the power of limits, and are thus called the "Supreme Active Powers". GHA DHA DHASH marks the forming of the finite through mass, physical/relative laws, mortality, and the power of sexual reproduction. Yet it is still all formed and connected through the toroidal wave of the One.

GHADHADHASH is the 9th Class of Sound that completes the foundation of the active powers of the quantum code in creation. This meaning is also found in the Sanskrit words *DHASH* and *DHA*, which derive from the root word *DHAtu*.

The Dhatus are the foundational building blocks of Sanskrit, a set of 2012 words in Sanskrit Grammar that represent *any idea*. Every word in Sanskrit is built from joining together (*sandhi*) these 2012 Dhatus. If you know the meanings of the Dhatus, you can derive, compound and create any Sanskrit word, because every Sanskrit word is derived from a Dhatu.

This is similar to how DNA instructions work, and it is no coincidence that DHA is used twice here, as DHA literally roots the foundation of the quantum code into creation.

The quantum or non-dual code always operates throughout all creation, yet now the new finite code for mortal life and death is introduced in the 9th step of GHADHADHASH. This new finite code flowers and fully manifests in the 10th geometric step of JABAGADADASH.

This marks the creation of the five sense organs, and the "beginning" of the temporary, physical world as perceived through the sense organs. This 10th step creates a circular geometric boundary around the 9, marking out the difference between the quantum code of the One and the establishing of a new finite code of cell mitosis and sequencing, which marks out the temporary and physical world.

GhadDhaDhash Practices

A simple, direct way to feel GHADHADHASH is by having a leg massage, a prostate/anus massage, and a yoni-lingam massage in the same massage session with a conscious Tantric practitioner. This can help you to feel and connect this energy flow consciously within you.

In sound, mudra and breath alone, we can also activate the flow *GhaDhaDhash*.

GHA: Legs and Feet

GHA is a guttural sound, made by contracting the throat and navel: just try and sound *GHA* in your *manipura* chakra, and you can feel the power in it that pushes. On your left hand, place your first finger and thumb together. On your right hand put the top of the thumb against the inside part of your ring finger nail. Focus on your feet. Sound *GHA* or *GHAM* for 3 minutes aloud, and 3 minutes silently.

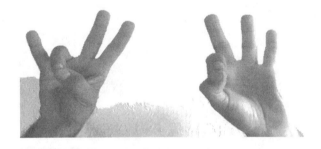

DHA: Excretion: Anus

DHA is a cerebral sound with breath in it, tongue pointing to the roof of the upper palate. Focus on your anus, inhale with *DHA*, contract the anus and exhale *DHAM*, releasing the anal muscles. Repeat. Find your own timing and rhythm in this for 15 minutes.

DHASH: Lingam: Procreation

DHA is a breathy dental sound. *DH* is sounded with the tip of the tongue striking the tip of the upper teeth, then opening with *AH* at the end of *DH*, to then come to *SHH. DHASH*!

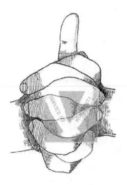

Do this mudra with *DHASH* for 5 minutes aloud, then 10 minutes silently. Focus on your lingam/yoni.

Tantric Practice

See your anus, your lingam/yoni, your navel and your legs connecting in a red triangle for a few minutes. Then start a sensual pelvic rocking motion, rocking your sacrum back and forth like you are making love, generating some sensual energy. Relax into this movement and rhythm. Whilst doing this, listen to some root chakra music with deep, mid-tempo rhythms.

Now add these sounds to the movements. **Say GHA**, movement, into manipura solar plexus chakra. Say **DHA**, release, into your anus. Breathe up energy at the same time from Earth through your anus. As you thrust your pelvis forward, say **DHASH**, focused on your yoni or lingam.

Release the anal muscles and allow the energy to flow forward out of lingam/yoni, and *back through* to anus.

Sound *GHA* into the spinal base, and then repeat.

Find your own rhythm with this, and be sensual in it.

One should feel joyful, energized, juicy and open if done correctly.

जबगडधश्

JA BA GA DA DASH:
The Five Senses

श्रोत्रत्वङ्नयन घ्राणजिह्वादीन्द्रयिपञ्चकम् ।
सर्वेषामपिजन्तूनामीरितं जबगडदश॥ २१ ॥

Śrōtratvaṅnayana ghrāṇa jihvād īndriya pañchakaṁ
sarvēṣām api jantūnāmīritaṁ jabagaḍadaś
21

Hearing, touching, seeing, smelling, tasting
are the five sense organs of knowledge
operating in all living beings
in harmony and order
as *jabagaḍadaś.*

The inadequacy of the senses to record the backward flow of forward moving things causes the illusion of sequence and time.
– Walter Russell

Our brain receives vibratory data and inputs from our environment, and then constructs our impression and vision of the external world by projecting outwards a hologram based on this data.

For example, our eyes interpret vibrations of light, which hit the back of the retina, triggering electrochemical signals that then travel to the back of the brain. The brain then, in less than one tenth of a second, puts this information together, saying this is what is happening "out there".

What an amazing apparatus! We pick up on and receive many different data streams all the time from the world around us, yet we usually detect and use only a small frequency range of this data, a mere 2 million bits of information out of a possible 4 billion bits of information.

What makes our brain select what data is relevant to us? There are several theories about this. One is that our brain picks out what our subconscious believes is important to our survival. Another theory is that the brain and subconscious pick out data from our environment based on our cultural preferences.

Another theory is that our brain and subconscious mind filter incoming information based on our past conditionings, our "filter bubble" of subjective personal experiences which creates an unheard, in the background, subconscious commentary and filtration system based on our memories, and what we have previously enjoyed and previously have not enjoyed.

We then believe all of this to be true, and what we believe to be true then reflects back to us our perception of the world. In this way, the brain and subconscious create a *reality tunnel*, a tiny window into an infinite flux of energy. All the senses help create part of your own reality tunnel, for as we relate to anything outside of ourself, we project onto it, creating a subject-object relationship. In Tantric terms, sattva or light is you as the subject, tamas or inertia is the object and rajas or action is the relating energy between them.

Each of the senses and their sense organs which create the sensory mind are media of communication, ingestors and projectors of data. In their interplay with the brain, they create a holographic vision of the environments and world around us. Part of the way we feel and think is influenced by what the senses ingest and project, and how the senses operate are influenced by what we feel and think as well!

The senses help us survive physically and protect us, enabling us to navigate through the three-dimensional world by making us perceive others as separate to us, and to perceive ourselves as an individual body-mind.

In this way, our sensory mind creates "particle-rization" of the universal wave-field, perceiving the ever-moving waves of creation as particles, things and objects. It is our sensory mind that makes particles and objects appear, for without the senses, no particles or objects would be perceived, and we would literally be swimming in an ocean of sound, light and wave flows.

When we do not have a sense of our internal *I AM or AHAM*, the senses become seen as the only reality and become the master of the Self. In Tantric terms, we become ensnared in *Maya*, the skin of the world, the appearances perceived through the sensory mind combined with our experiences and memories that lead our subconscious to create our unique version of reality solely *relative to us*.

This leads to delusion and the forgetting of AHAM if one believes that forms, sounds, smells, tastes and touch are the *only* reality. To perceive behind this skin whilst simultaneously enjoying it is part of Tantra, one definition of which is "extending the thread of the body to include all".

All the senses must be fully alive, active. To be sensitively aware of thought and feeling of the world around you and of nature is to explode from moment to moment in affection. Without affection, every action becomes burdensome and mechanical and leads to decay.
– Krishnamurti

The senses are *jnanendriyas* or organs of knowledge, so called because the evidence given to us by our senses is the first of the *Pramanas*, or means of *right knowledge*, which works through *pratyaksha*, meaning "before your very eyes". The senses in this context are the servants of the Self, not the other way around as is seen in modern society.

The senses are receptive powers for Self-recognition, helping us to recognise the Self in all life. They are instruments and conduits of knowledge, designed as our servants to help us perceive the living light of sentience in every sentient being. They can serve us to enjoy the Self in all things, and by using them as interfaces to travel beyond them, enable us to experience the infinite within every form.

Perceiving clearly through the senses happens when the conditioned ego-mind *ahamkar* is stilled, if even for one moment, allowing our lifetime's worth of conditioning about what it is we are seeing, hearing, tasting, touching and hearing to be forgotten, leaving presence, openness and aliveness to feel exquisite sensations.

Reality is appreciated through the senses when the mind stops interpreting what it is experiencing, ceasing its subconscious narrative commentary on what it is experiencing, instead becoming present to what is here in the present moment, in the power of Now.

This is celebrated in Tantra, which includes the senses but does not rely on them, as Tantra is the ability to engage with anything and everything without judgment. To be totally engaged also brings total detachment; total absence brings total presence, and total engagement with everything allows one to feel and see the Self everywhere.

The highest purposes of the senses are for us to enjoy life, to appreciate it, to derive pleasure from it and all that the living light, which is in all of existence, offers us by perceiving the literal wonders of the world. Appreciation brings this forth palpably, and the vehicle of the senses helps us see, hear, feel, touch and taste this living light in all beings.

Through our eyes we can see the light, auric fields, sensational colours and glowing life-force of plants, trees and people, drinking these in and enjoying the beauty of the world. Visual beauty can point towards the eternal if we see it like this. Through smells, we can be healed of all manner of ailments, be uplifted and energized, returned to our most cherished memories, and appreciate our primordial, earthy self.

Through taste we can enjoy sensual pleasures and imbibe the ambrosia of life's great foods and elixirs to nourish and vivify ourselves. Through touch, we feel connected to each other, to affection, to being held, to love, and to all life. Through hearing we can enjoy the most sublime of emotions and enter space and beyond into our origin of silence, the heart of all creation.

The senses are a tool for enjoying the living light in everything: this pleasure is the delight of tantric artists, where all creation is embraced, celebrated and enjoyed as part of the dance of Shiva and Shakti, as part of their lovemaking that creates each of us and all worlds.

In the *Vigyan Bhairav Tantra*, the discourse given by Shiva to Shakti revealing 112 ways to meditate, Shiva shares more about the senses:

All these beautiful surfaces decorate vibrant emptiness. Every perception is an invitation into revelation. Hearing, smelling, tasting, seeing and touching are ways of knowing creation, for the deepest reality is always right here. Follow your senses to their end and beyond, into space. [154]

All around you, in every moment, the world is offering a feast for your senses. Attend this banquet with loving focus, as the outer and inner worlds open to each other. Become identical with the ecstatic essence embracing both worlds.[155]

The Vedic sages too enjoyed the five senses, as they were known as the *indriyas*, companions of Indra, King of the Gods, who delighted in being intoxicated on *Soma*, the juice, fuel and blissful elixir of life.

Vision is food for the eyes. Sound is food for the ears. Touch is food for the skin. Taste is food for the tongue.
– Dr Vasant Lad

We are influenced by what we take in through the senses, for we literally absorb and ingest these stimuli: we take them into ourselves, they become part of us, and we react to them. We can enjoy this, or not, all of which triggers other physical, emotional or mental responses.

In one sense, we are what we eat, as what we eat literally becomes part of our body through the tastes we enjoy *and* the tastes we leave out, all of which influence our moods. We are influenced by what we see as we drink in this data, and our bodies merge with what we touch, taste and allow into us.

What we touch, what we are touched by and what we allow to touch us influences us. What we smell influences our subconscious brain and can trigger emotional memories and traumas, and the smells we find appealing, healing, sensual, disgusting and pleasurable stimulate our primal brain to evoke and form memories *and* inform us of primal emotional conditions.

The frequency and harmony (or not) of the music we listen to, the sounds we make and hear and the ambient sounds surrounding us at all times contribute to our sonic diet and state of well-being. All these sensory experiences help your mind to process and interact with your surroundings, and leave physical, mental and emotional impressions that shape part of who we think we are.

To lend another spin to this, the senses all connect to each other, as seen through the phenomenon of synaesthesia, where certain people can see sounds, taste colours and hear forms. The senses become cross-pollinating. For example, sight and sound are correlated, as University of Oregon neuroscientists Bala and Takahashi discovered when studying owls. Every time the owls heard an unexpected sound, their eyes dilated.

Excited by this discovery, they experimented on how to use the eyes as a window to hearing, and found similar involuntary dilation in humans, eventually publishing their findings in the *Journal of the Association for Research in Otolaryngology*, where they used eye-tracking technology in traditional hearing exams.

This cross-connecting of the senses opens an interesting set of questions: Without being able to hear, would you believe that sound exists? How would your sense of space differ? Without being able to feel touch, would you feel as connected to yourself and others as you do now? Without being able to see, how would you react? Without being able to taste or smell, would you enjoy or even want food?

We all have our own unique smell that identifies us to others by the pheromones behind our ears, genitals and armpits, which hold our unique scents or olfactory signatures. This is similar to the sonic signatures held in our electromagnetic fields, our unique frequencies.

Our eyes hold our unique signature through iridology lines marking out the state of our organs' health, traumas and emotions; our fingers have unique fingerprints, as do the lines on our palms which a Vedic astrologer versed in *Jyotish*, the art and science of divination, can decipher.

Whatever sense we are most comfortable with, and use the most for navigating in the world, is what we most identify with as our primary element. If we call ourselves a "sound person", space element is our home, and sound is a nutrient to our nervous system, acting like a natural pharmacy. If we are more of a visual person, enjoying forms and colours is what we most enjoy and we are more fiery. If we most identify with smell, we are a more earthy person, and if we enjoy taste the most, we are more watery.

However, most of us identify as being primarily sonic or visual people, for these are our two main navigating tools: to see and to hear is how we judge others and the world.

JaBaGaDaDash

*Man is an antenna, receiving and transmitting music that gives
life to the world.*
– Alfred Tomatis

An aspect of this music of life is translated through our senses,
as all our sense organs receive vibrations. For example, our
bodies are wired for sound, and your ears interpret vibrations
of sound, as Sufi Master Hazrat Inayat Khan shares:

*A person does not hear sound only through the ears; he hears
sound through every pore of the body. It permeates the entire
being, and according to its particular influence, either slows
or quickens the rhythm of blood circulation, either awakens or
soothes the nervous system.*

The skin and nerves interpret vibrations through the physical
body, but the skin also perceives inaudible subtle vibrations
from the air. People "in touch" with this are able to perceive a
storm before it comes, or sense danger before it arrives, as the
skin is a fine sensing organ which contains dermatomes, mini
Wi-Fi-like transmitters which send information directly to the
spine and vibrate the organs of the body.

Our bodies and senses are giant receivers: the eyes take
in light, the ears take in sound, the skin interprets ultrasonic
vibrations. The brain is like a TV set in this process, acting as
an intermediary translating different wavelengths of vibration
each of which produces different images, sounds, feelings and
experiences. If your inner TV set is tuned into the senses alone,
without consciousness of the previous classes of sound and
functions of the Self, then you will only access a few channels
of information.

In *jabagadadash* comes a culmination and coalescing into form of the five great elemental powers *mahabhutas*, their measures *tanmatras*, the *karmendriyas* or active energies of Self-expression, and now the five sense organs *jnanendriyas*, the receptive powers of Self-recognition.[156] All are formed to express the Self.

Interestingly, the word "jabber" is said to come from *jabagadadash*, the syllables for the five senses. Its origin is said to come from the time when the British invaded India. In an outpost in the Rajasthan desert, a Tantric school was practising the Maheshwara Sutra.[157] Twenty or so initiates were chanting the sounds at different timings to produce an incantatory, entrancing effect. As the British officers stormed in, they were amazed to hear what sounded to them like a cacophony, a babble of sound.

As the story goes, the phrase they heard was "jabagadadash" and from this the modern term "jabber" arose, bringing to mind a cacophony of cascading streams of chaotic noise, like a radio constantly babbling or a person jabbering away at you.

This is an apt term to describe a situation when the senses are ill functioning through being overused, underused and not being used at all. These three conditions are seen in Ayurveda[158] as being *one of the three main causes of disease*, the other two being time and ignorance.

Ja: Srotra – Hearing and the Ear

JA is the seed sound of *Srotra*, the receptive power of hearing. The ear is the *only* external point of emergence of the vagus nerve, the longest of the cranial nerves, which travels from the brainstem to almost every major organ in the body. This nerve influences our biological rhythms, is part of the autonomic nervous system, respiration, cardiac and circadian rhythms, as well as supplying energy to our vital organs.

The vagus nerve is directly connected to our ears, and vibrations entering the ear send information to the spinal cord, the brain and almost every major organ. The ears are linked to both hemispheres of the brain, and inside the ear lies the organ of Corti, a spiral shaped receptor organ that is both a receiver and transmitter, transducing auditory signals into nerve impulses, transforming stimuli from our environment into usable vibrational energy.

The ear is the second sensory organ to develop in the foetus after the tongue, and the bones of the inner ear are fully formed before birth, the only bones to be so. The petrous bone of the inner ear also contains a high quantity of DNA, allowing scientists to use genetic material from old skeletons.

The ear never stops working: even when asleep, our ears are alert. Every sound we make is dependent upon our ability to hear the sound (the sounds we make reflect the sounds we hear). No other sense organ can register impulses as minimal as the ear can, for the ears are more accurate than the eyes and can even perceive numerical quantity!

Listening and hearing are different functions. Hearing is the passive ability to receive audio information. Listening involves filtering sounds, analysing the tones and feeling behind them, and then responding. Listening is an active, refined skill that encourages the brain to change, which helps other biological systems to reorganize. This is similar to the principle that singing and playing music activates more parts of the brain than just listening to music.

Reorganizing the brain by altering your auditory inputs can improve reading skills and improve communication, attention span, reading comprehension, speech quality, memory and spelling. The more you listen, the more you learn, with your brain reorganizing itself to do this.

Space, sound, hearing and speaking are the four links in the chain of the manifestation of sound into your world. Without being able to hear, you lose your sense of space, you lose your sense of orientation, your sense of navigation and your voice. You live in a sonic void. Hearing loss affects our ability to connect to the space element, as space is the ground of all elements and the medium for matter.

Most deaf people think in the images and pictures of sign language, which then becomes the medium of their inner voice. Sign language, whilst useful, has a conditioning effect on their thoughts. Deaf people were once thought of as unteachable (thankfully, this is no longer the case) until they could learn a language (sound works internally too in thought!).

Quiet people have the loudest minds, according to scientist Stephen Hawking. The mind tries to compensate for the lack of vocal expression with thoughts and formulations, analysis and introspection. When they think, they have more images (fire element) rather than hearing a voice as most people do.

Sounding J

This is a palatal *ghosa* sound, as in "jar" or "jam". J is made with the tip of the tongue just touching the lower teeth at their base, with the mouth open.

Ba: Tvak – Touch and Skin

The skin is the one goal of all kinds of touch.
– Brhadaranyaka Upanishad 2:4.11

B is the seed sound of *Tvak*, the power to recognise the Self through feeling and touch. The sense of touch pervades all other senses, and without the sense of touch the senses cannot make any sense.[159]

Touch is felt through our skin, which covers our entire body and has its own ecosystem. The skin contains a network of tiny, egg-shaped pressure centres called *Pacinian corpuscles* that sense touch, and are in contact with the brain through the vagus nerve, which is connected to oxytocin or happiness receptors.

Stimulation of the vagus nerve triggers an increase in oxytocin and a cascade of wellness, happiness and health benefits.[160] As we have seen previously, the vagus nerve travels from the brainstem to connect to almost every major organ in the body, feeding, nourishing and informing.

The skin is a transceiver for the vagus nerve through dermatomes, zones of skin whose sensory input is connected to individual spinal nerves, creating direct highways of information and sensation between the central nervous system, brain and skin. Dermatomes roll around our entire body, so our skin and sense of touch are direct links to our brains and spinal cords, which connect to every organ in the body.

Scientists believe that touch is the first sense to develop in utero, and therefore touch is crucial to our early development as children. The more we are touched, held and hugged as children the more our brains, cognitive skills and emotional intelligence develop. Touch is important to relationships, intimacy and team building. Without touch, emotional connection and self-love can diminish, along with an aspect of our humanity and ability to connect, relate, and to be touched: to feel alive.

Without touch you can feel isolated, alienated and disconnected to life, to yourself, and to others. Touch and emotion are so intertwined that tactile information can cease to make sense without emotional meaning. We speak about being "in touch" with our emotions and being in touch with ourself. Yet, a blind person can learn to discern colour by touch and a deaf person can be trained to "feel" the vibrations of speech through touch and skin.

Sounding B

This is a labial or lip-focused sound, made with the tongue relaxed and lips pursed together. *B*!

Ga: Nayana – Sight and Eyes

G is the seed sound of *Nayana*, the power to recognise the Self through sight. Nayana literally means leading, as it leads all the other senses in navigating through the world. In seeing, light enters the eyes, hits the back of the retina and triggers electrochemical signals which travel to the back of the brain. The brain then, in less than one tenth of a second, puts this information together, and then says this is what is happening "out there". This reality tunnel is partially formed by what we want to see, subconsciously filtering out that which is deemed to be irrelevant.

Form and colour are perceptions provided by Fire element. The eyes are vehicles through which light is ingested, yet colour is not found in light; colour is created by neural processing circuits in the visual pathway and brain.

Problems with the eyes are connected to the fire element. Training your vision to improve does not alter the photoreceptors in your eyes. While all the same sensory information is getting into the system through these receptors, specific training *allows the brain* to filter out noise and more effectively "tune into" the sensory signal.

So, imagine: if you could not see, what would you be left with? You would come to rely more on the other senses, which would become heightened and sharpened in your inner darkness. Your sense of hearing may become more acute, your sense of touch may stretch out to feel non-physical energy fields, and your sense of taste may become refined and exquisitely sensitive to the slightest quiver. Your sense of smell may also become more acute, enabling you to perceive danger, safety, disgust and arousal more easily.

Many people today are experimenting with the senses. One ancient practice connected to cutting off your sense of sight is to live in total darkness for extended periods of time in order to expand one's consciousness and develop the subtle senses. In these "darkness retreats", people are blindfolded for three or more days in specially prepared blacked-out houses to enter these altered states of consciousness.[161]

Darkness is the state we dream in. From darkness, all light arises. Darkness is where the universe sprang from and the place we are nurtured in during the first nine months of our life – in the womb. Simply to sit in darkness allows us to feel beyond ourselves and who we think we are. We subtly start to feel more things that we are usually unaware of, brush off, or dismiss.

When we are in darkness together, we become closer. In darkness retreats in Thailand with Dr Mantak Chia, and in the UK with Simon Buxton, they found that without the bonds and constraints of what we look like through gender, age and race, we become closer to each other and more equal. In this context, *it could be said that light divides, and darkness unites.*

The sense of sight, the tanmatra of form, the element of fire and the eyes form one chain of consciousness. Yet with light and sight comes the reminder of its source: darkness.

Sounding G

This is a guttural, powerful sound made with mouth open, tongue relaxed, with a contraction at the back of the throat (say gargle). It uses minimum breath.

Da: Ghrana – Nose and Smell

D is the seed sound of the organ that cognises the Self through smell or *Ghrana*. The nose is designed to smell vibrations, and all smells can be measured in terms of frequency: for example, rose oil has one of the highest frequencies of all essential oils.

This is similar to the science of psychoacoustics, which measures the Hz frequency of all objects, such as the organs in your body, the planet earth, the sun, water and more. Everything has a frequency, and together they can be correlated, so every image has a smell frequency, a colour frequency and a sound frequency.

Olfactory information is carried through odour molecules, which are represented in the brain as *odour maps or odour images*. The image patterns for different odours are different for different molecules, and are formed within the olfactory bulb, which forms "odour objects" in the olfactory cortex, which are then integrated by the brain into odour perception.

Smells trigger memories and feelings: memory triggered by smell is more long-lasting than any other kind of memory, which is why childhood joys and memories are so easily evoked through smell, such as smelling your mother's homecooked food. Smell is more potent at triggering a memory than any other sense, and can evoke empathy, influence our moods, influence our choice of sexual partner, as well as stimulating the endocrine system and its associated emotional triggers.

Smelling helps us detect changes in our social atmospheres and interactions. Without having a sense of smell, an ocean of past images and emotions disappears from your memory, and this can be felt as a loss and a grief. Smell can change your mood and behaviour, as essential oils and male-female pheromones testify to, and certain odours help our immune system.

Smell can signal danger and death, and is a means of protection. A fading sense of smell experienced by the elderly means death is coming closer. Smell protects you from illness and danger, ensuring that you realize your food has rotted, that the gas is on, and alerts you to the smell of fire burning. Smelling fear coming from others is a signal for a predator to attack; if you have no fear, there is less likelihood of you being attacked by a predator.

Losing your sense of smell (and by association taste[162]) can come from deep grief, trauma, disturbance of the lung meridian, brain shock, viruses and nasal issues. When people lose the pleasure of smell, it can take away part of their joy of life, with other senses such as sight becoming heightened as a substitute, as what you see can help your mood.

Some people who have no sense of smell and therefore no taste have trouble identifying the emotional states of others, as they rely more on appearances for information, which can often be deceiving. Without smell, you lose interest in food and taste, and the sensual aspects of life become dimmer. You lose weight, connected to the earth element of your body (as smell is part of the earth element's chain of manifestation), which then affects the organs of procreation and your urge to reproduce.

Sounding D

This is a cerebral, minimum breath guttural sound, made with mouth open and tongue tip striking the midpoint of the upper palate.

Dash: Jihva – Tongue and Taste

Eating is the only thing we do that involves all the senses. I don't think we realize just how much influence the senses have on the way that we process information from mouth to brain.
– Award-winning chef Heston Blumenthal

The seed sound **Dash** generates *Jihva*, the organ that recognises and receives the Self through tasting, which now completes the water element in manifestation. As with all the senses, taste is not in food; it is created by the brain through multiple sensory, motor and central behavioural systems, including the respiratory, cognitive, emotional, language, pre-and post-ingestive, hormonal and metabolic systems.

In the body, taste information is transmitted to the medulla, thalamus, limbic system and the gustatory cortex tucked underneath the frontal and temporal lobes.[163] These centres are related to learning, perception, the subconscious and primal self, the super conscious of the thalamus, the sense of the individual self or *ahamkar*, and to emotion and memory.

This is quite astonishing in its complexity and interconnection! More human systems are engaged in producing taste perceptions *than in any other human sense behaviour.*

Up to 70% of taste is due to smell that occurs when we are *breathing out.* Sound is also a factor affecting taste, and top chefs are now making their dishes sonically interesting, using everything from a sprinkling of popping candy and crunch, through to iPods playing the sound of waves, such as at chef Heston Blumenthal's award-winning Fat Duck restaurant.

We are largely unaware of the many subconscious contributions to taste. The conscious sensations of the basic tastes from the tongue "capture" our awareness of the food and refer all other sensations, including smell, to the mouth. Flavour therefore has the quality of an *illusion*.[164]

This makes the idea of flavour vulnerable to manipulation, as is well recognised by marketers who spend millions to influence what we eat and drink, selling an idea around who we will become and what lifestyle and social clique we are buying into if we eat or drink their product.

Flavour is *multi-motor*, involving the tongue, jaw and cheeks. The movements of the tongue in manipulating food in the mouth are more complex than the movements used in speaking English, yet this is different for the sounds of Sanskrit which use the tongue and mouth more extensively.

Taste is *multisensory* in nature: it is deeply influenced by smell (arising from earth element) and both have sensory receptors that respond to molecules in what we taste and smell. When we describe the taste of any food, we are actually referring to both the taste-gustatory and smell-olfactory properties of the food *working together*.

As famous French chef Paul Bocuse said, "the ideal wine... satisfies perfectly all five senses: vision by its colour; smell by its bouquet; touch by its freshness; taste by its flavour; and hearing by its *glou-glou*."

As all good chefs know, there are seven tastes: bitter, salty, sour, astringent, sweet, pungent and umami (deeply savoury). Yet, how a person perceives taste also has to do with the classic twofold conditioning factors of nature and nurture.

"Taste is a product of our genes and our environment," says Leslie Stein PhD, from the Monell Chemical Senses Center in Philadelphia. "DNA plays a part by giving a person taste preferences. And our environment is a factor in learning new tastes. Part of why you might like broccoli while your best friend finds it bitter is because you have different genes, which code for different bitter receptors."

The water element completes its chain of manifestation in taste, having flowed through the organs of evacuation, the nourishing power of consciousness, your emotions and your sensual, creative, fluid self – found in your *swadhisthana* sacral chakra.

Sounding D

This is a minimum breath, dental sound. With mouth open, sound D by using the tip of your tongue to hit the lower edge of your top teeth. *Dhash* is pronounced with a swift "ush" after the DH.

फछठठथचटटव्

KHPHCHTHTHCATATAU:
The Pranic Mind

प्राणादिपञ्चकं चैव मनो बुद्धिरिहङ्कृतिः।
बभूव करणत्वेन खफछठठथचटटव् ॥ २२ ॥

prāṇādi pañchakaṁ chaiva manōbuddhirahaṁkṛtiḥ
babhūva karaṇatvēna khaphachaṭhathacaṭatav
22

The flow of the five pranas sustain mind, intellect and ego,
and evolve through the process of
khaphachaṭhathacaṭatav.

वर्गद्वितीयवर्णोत्था: प्राणाद्या: पञ्च वायव: ।
मध्यवर्गत्रयाज्जाता अन्त:करणवृत्तय: ॥ २३ ॥

vargadvitīyavarṇōtthāḥ prāṇādyāḥ pañcha vāyavaḥ
madhyavargatrayājjātā antaḥkaraṇavṛttayaḥ
23

The five winds of prāṇā
arise out of the semi-vowel in each group *(h)*

The consonants are prāṇā and the five winds.

The subtle moving thought flows
of the inner perceiving organ *antahkarana*
emerge from the first letter of the three middle
groups (*ca, ṭa, ta*).

In the Vedas, Brahma breathes creation into existence through Prana. *"Prana is Paramatma, Antaratma, the Supreme Being beyond and in bodies as their controller and director. It pervades and vitalises all living creatures as the source of the individual and collective life."*[165]

Prana animates all living things. *Prana* is more than breathing, but includes the breath. Prana is in our breath and beyond it. Oxygen is part of Prana, but Prana is far more encompassing. Prana is the primary vital energy or vital force of life, also known as chi.

Prana is the energy of life itself, not just our human life but the universal life-force. Prana powers the movements of the universe. Prana powers all our cells, senses and mind; without it they cannot function. Without Prana, there is nothing we can do and no energy with which to do it!

There is only one Prana, yet it flows in multiple ways. The universal prana flows through our body-minds in five winds, airs or flows known as *Vayus*, which fuel and are organized by the trinity of the mind: the separate ego-self *ahamkar*, the data-mind of *manas*, and the conscious, discerning *Buddhi* intellect.

All five prana flows connect with each other inside you, energizing, fuelling and sustaining your body and mind. The most sustainable foundation of spiritual health, mindfulness and well-being is having these five pranic flows open and moving freely. These five are: *prana, apana, vyana, udana* and *samana*: life-force, release, circulation, ascending and balancing currents of prana.

These energy flows travel up, down, around, in and out of us all the time, even whilst you are reading this. When one of these flows becomes imbalanced, so do the others because they are all interconnected.

The root word or *dhatu* for Prana is *AN*, meaning animating, breathing, vivifying, enlivening. The different prefixes to *AN* show the four other flows of prana: *apana, samana, vyana* and *udana*.[166]

Prana is in your inhalation, moving energy from the heart to the navel, *the* Pranic centre in the body. It brings in energy and light on all levels, governing the intake of all substances. This Prana feels the best for us as it energizes, activates, uplifts and expands our energy. It moves in and up.

Apana is its twin, found in your exhalation, releasing toxins, wastes and baggage, ranging from carbon dioxide to many other physical, emotional and mental wastes. Apana releases that which we no longer need and that which is bad for us, that which weighs us down, that which is judgmental and negative. It moves energy from the navel down to the anus and out.

Samana governs the absorption of oxygen, moving energy from the navel to the entire body. It converts fuel to energy and is fire, governing the digestion of everything we take in, be it food, emotions, ideas or experiences.

Vyana governs circulation, moving energy throughout the entire body-mind, circulating energy to where we need it, keeping the body-mind running smoothly and in harmony so all parts of us can connect and communicate with each other.

Udana governs the positive ascending energy up through the throat into the brain. Udana is connected to sound and consciousness, and is the most "spiritual" pranic flow.

The five prana flows first ignited in our bodies when we were foetuses, carving channels in our tiny bodies and brains to allow energy to flow and form the structures of the mind, energizing them into manifestation. As the pranas unfurl and leave our bodies during the death process, our consciousness is transported into other dimensions. Many yogis and lamas use the five pranas and specifically Udana, in order to consciously leave their bodies at a time of their own choosing.

The five currents of prana energize and help form the structures of our mind. When our minds are out of balance, stressed or worrying, when we are more reactive and angry, it is because the five pranas are out of balance. When the pranas come into balance, mind comes into balance.

The mind is fuelled by the pranas. When the pranas are flowing well in balance, our ego is in balance, not running the show, not being the master of our hearts but rather being the servant of our hearts. When our pranic flows are out of balance, the ego mind runs amok, running our lives and creating misery for everyone.

The more you clear and amplify the pranic flows, the more energy you have available. However, if a pranic circuit is misaligned or weak, energy leaks and drains away. You don't have enough energy. The mind then becomes restless and fidgety; it becomes harder to concentrate, focus and be mindful.

One cannot digest life's impacts, finding it hard to release stress, toxicity and negativity. Energy does not flow optimally to connect up the different parts of you. You can feel stuck, without the fire to move through the challenges and opportunities of life, to get up and go, to break through blocks. Your body can feel heavy and unresponsive. Your mind can feel lethargic and dull.

The five pranas are the fuel for the mind. If there is not enough fuel, the mind will be weak, foggy, easily led and unable to think for itself. Balancing our pranas makes our minds clear, sharp and calmly active, focused, relaxed and aware.

The body cannot live without the mind "for mind and prana are twins in function". The five pranas are the energy source for the mind and its distributor, "connecting with the body and the universal energy supplies of the four elements." Together, the mind and pranas create a "wireless circuitry that interconnects every point of the subtle form with every other point in the energetic system."[167]

The five pranas can be activated through yoga, breath, mudra and mantras that contain the seed sounds of the pranas themselves. Each of us has one prana that is weaker and needs amplifying, and one prana that is our strongest. Identify which prana flow is your weakest, and work more with this one.

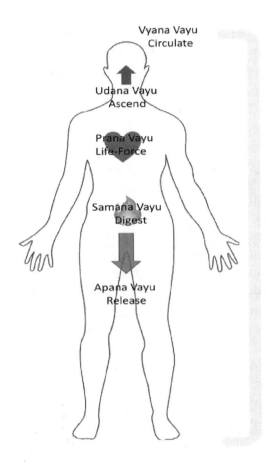

Pranic History

There are different stories in Indian spiritual literature, specifically the Upanishads, about the five pranas arguing with each other as to which one of them is the most important. To resolve their dispute, they decided that each of them, in turn, would leave the body and see whose absence was most missed.

As Prana left, the body began to wilt and die as there was no life-force left to power it! All the other pranas then begged it to return, honouring Prana as the supreme power of the body-mind.

A modern technological version of this conversation lies between Facebook saying, "I know everyone," to which Wikipedia replies, "I know everything"; Google retorts, "I have everything"; to which the Internet scoffs and says, "Nothing works without me." Electricity (prana) chimes in: "Oh really!"

In another (ancient!) version of the story, it is apana who is acknowledged as the most important pranic flow, as nothing could leave, exit or release from the body when apana left: the body stagnated, toxified, became totally blocked up, and started to die too! In each version of the story, it is prana or apana who wins the argument.

> *All that exists in the three heavens rests in the control of Prana. As a mother, her children, oh Prana, protect us and give us splendour and wisdom.*
> – Prashna Upanishad II:13

In Vedic literature, the Pranas are called Rudras in the Brihadaranyaka Upanishad (III:9:04), and the Rig Veda clarifies that the Rudras (also called the Maruts) are the children of Rudra-Shiva. The Pranas and their attendant functions of the mind, intellect and ego are the children and manifestations of Shiva, the One.

The Vitality Body

The five flows of prana create *our vitality body which fuels our mind, intellect and ego.* Our vitality body connects, fuels and runs our body-mind. Our earthly life is largely dominated by our vital body and our need for energy.

Our vital body contains the deep-seated urges that keep us alive, such as survival, sex and movement. It is part of the subconscious that holds our fears, desires and attachments, powering our need to seek enjoyment through our senses and mind.

To create changes within us we must understand the energy through which things work, and then do something to change it. Prana flows where attention goes. A person with a strong vital body is powerful, prominent and able to impress themselves upon other people and the world at large.

These people get things done and motivate others. With a strong vital body, *we have the power to do anything* with energy, curiosity and enthusiasm if we harness this energy effectively. A strong vital body becomes magnificent if it is directed beyond ego desires: into the heart.

Those with a weak vital body lack power and are more passive: they are followers rather than leaders. They do not accomplish much, have little forward movement in their life, remain in a subordinate position to others and are victims to the eddies and currents of life.

People with strong vital bodies chart their own course in life, always moving forwards and impressing themselves upon others. They are masters of their own destinies, and run the world: people with weak vital bodies follow them.

When your vitality body is strong and clear, you are naturally "protected" to a large degree from negative environments, pollution and negative energies, such as toxic people, EMFs, 5G frequencies, disease, illness, viruses and environmental pollutants. This is because the five prana flows create a circuit that inhabits itself *within you and around your body*, providing an external buffer to reinforce your body-mind.

Let us now dive deeper into these five flows of our health, energy and well-being: the energies of increasing our life-force, releasing all that does not serve us, circulating and communicating information, energy and nourishment to all parts of us, having our ascending spiritual energies flowing, being able to digest life, and being able to change ourself: Vayu Yoga.

Prana

Since Prana is the life of all beings, therefore it is called the life of all.[168]

Pra means the first and foremost, and one of its main sources for us lies in our inhaling breath focused at the heart. Prana is an inward and forward moving current governed by the air element.

Without breath, there is no life. The rhythms of our breathing flow and tune our body-minds to the waves of vibration that flow through space. Breath runs throughout all life as a wave, moving through the body-mind like a wave moves through an ocean, changing *and* connecting all it touches.

It has been said that if we were to change nothing (not our eating, exercise, thinking patterns or habits) except our breathing, we could radically alter our life. Breath is life and life is breath.

As we inhale, spiralling rhythms of breath radiate outwards from every cell and atom; as we exhale, we spiral inwards. These spiralling currents of life-force undulate through five phases of resonance – an inhale, a middle breath, a diffused breath, an exhale, and an upwards breath.[169] These course through our heads, throat, heart, belly and root: through every part of us.

With every breath, we breathe in 10^{23} atoms – a thousand trillion-trillion atoms, which join us and become part of us; we then breathe out a similar number of atoms that were, just a moment before, part of our body-minds. Through breathing, we are in this constant state of transforming and exchanging life-force with the rest of the universe *all the time*: we literally could be breathing Buddha's or Christ's air at this very moment!

Through deep breathing or *pranayama* we align and attune all parts of our body-minds, for breath is the bridge between conscious and subconscious. Breathwork can powerfully heal our body-minds and nervous systems, releasing old memories, traumas, beliefs, stresses and emotions, simply by more life-force flowing into all parts of your body-mind. Through deep breathing we vitalize and balance our body-mind, and access much more energy to engage in life and be here now.

Breathing deeply makes us feel good! It brings us more energy, and can lead to healing, peace, centring, relaxation and deeper meditation. Breathing exercises or *pranayama* relax, release, open, and control the mind and senses as they take awareness inward into a more refined prana.

On another level, when we breathe in we are drawing breath from the universe which is breathing out into us. Mystics experience this as God breathing them: they are being breathed by God. Our relationship to trees and plants is a facet of this as well, in the exchange of carbon dioxide and oxygen which benefits us both.

Prana is not just in the breath, however. Pranic levels within us are boosted when we take in more sunlight, more fresh air, when we breathe deeply, drink good quality water, spend time in nature, have a good amount of sleep, exercise, meditate regularly, and have positive relationships. A fresh, organic vegetarian diet is rich in Prana, as living foods full of life-force support Prana.

Prana is the flow of life-force that enables you to move and think, your fundamental energizing force, *the* energy that sets things in motion, inspiring movement, forward energy and motivation. We bring in Pranic energies and information from the outside world, in everything from air and food to impressions, ideas, nourishment, feelings and knowledge. In this process, we are continuously exchanging energies from outside ourselves, bringing energies in and releasing energies out.

Prana organizes the five senses and allows us to receive through our senses. Prana is the fuel for our senses to function and bring in clear impressions from our environment. Prana brings heightened sensitivity to our senses, making our eyes see clearer, our noses to smell more acutely, our ears to discern sounds more accurately, our skin to become more sensitive, and our taste buds to become more refined.

If prana is unbalanced we suffer from needy cravings, become misdirected and off balance, develop bad habits, have less energy and become restless, dissipated and scattered. With more prana, we feel alive, move freely, and have a calm mind that enables us to respond to life's changing circumstances. *We can be more in the moment.*

Prana helps us develop mindfulness and well-being. It helps us see people and the world positively and to be inwardly content, *when it is balanced.* Discerning and taking in positive, life-affirming impressions from your environments and from other people support high Prana levels. Having good relationships is a good source of Prana. Some people we feel elevated with, others we feel drained by. This is all Prana!

Meditation alleviates the need for a lot of sleep and food and connects you to higher frequency Prana. Meditation creates space in the mind and unfolds more Prana. When the mind is quieter and more receptive, new energy can come into it. Prana works well with sound. The vibrations of the mind follow the vibration of Prana: what you focus your attention and prana on, you become.

Unconditional love is the flowering of prana in the heart. Selfless acts of love and feelings of unconditional love generate prana that leads into the Supreme Prana.

The Deepest Meaning of Pranayama

Paradoxically, the Supreme Prana and the infinite is accessed by *the ceasing of our breath*. This is the deeper meaning of the word prana-yama, which literally means the death of breath. This death of breath is the goal of all yogas, and occurs through the development and control of prana *and* breath that can then lead one into the bliss soaked *nirvana* of Samadhi.

This occurs as the breath slows down to a breath or two per minute for hours at a time, releasing hormones from the pineal gland and taking one into a state of absorbed, one pointed, deep . bliss. *This is the ultimate purpose of pranayama*: not just to practise for the sake of health, healing and well-being for the sake of the body-mind, but to bring you into the supreme prana of nirvana and the infinite through the death of your breath.

Most breathworkers and healers use the breath to facilitate somatic and emotional healing, to increase the amount of energy one has, to expand one's sense of well-being, to advance one's yoga practice, and more. This is all useful, and deeper breathing to gain more prana is necessary up to a certain point *as a preparation and purification* for the body-mind to let go of itself and enter higher states of consciousness beyond body-mind identification.

However, the supreme prana is beyond breathing. It is truly accessed when we no longer rely on breath to facilitate our sense of self. All five pranas feed and create the trinity of the mind, meaning that when these pranas shift in relationship to the mind, *the mind and sense of being a separate self can cease to be.*

In the uplifting and refining of the pranic flows we access different state of consciousness, as you may have experienced in breathwork sessions. These states are a preparation for the body-mind's energy flows to let go of its identification to the mind and the individual self, and to let go into something far greater. In transcending the limits of the body-mind system, we access the realm of the supreme prana.

The soul is not dependent on the body-mind to exist and needs no breath, sleep or food, as yogis have shown us for thousands of years. The death of breath leads to one directly experiencing, viscerally and tangibly, in every cell of your body, the fear of dying. If you are identified with the body-mind you will have this fear, and this fear will assuredly arise when your breath and heartbeat slows down to almost nothing.

Prana Yama invites you into this experience so you may fully embrace your fear of death, and enter That which is beyond the body-mind and death of the physical self: the eternal soul, the One, pure awareness independent of all things.

This conquering of the fear of death is actually the beginning of the Tantric Path, not the end. Conquering the fear of death is the beginning of the end of all your illusions, attachments and sufferings. You may, right now, intellectually believe that the eternal is within you: but you will only Realize this directly and experientially when you confront the fear of death and overcome it, thus beginning to live the timeless truth of who you really are.

Specific training and practices can lead you into this direct experience if you want to awaken into the eternal, beyond your body-mind identity, into freedom. This then is the height of all yogas, the goal of meditation, the ultimate stage that all breathwork and all spiritual practices prepare you for.

Simple Prana Practice

1. **Do Breath of Fire for 3 minutes in Gyan Mudra.**
   Thumb connects to first finger. Rapidly pump the navel in and out whilst breathing through the nose.
2. **Focus on your heart** with your hands in Prana Mudra. Place your first two fingers straight, little finger and ring finger connecting to the tip of your thumb. Sound the seed syllable of Prana, **KHA or KH**, for 5 minutes or longer.
   To sound KH: Have the mouth wide open, tongue relaxed, with a slight contraction in the throat to make the semi-guttural sound *KH*.

3. Whilst still in the mudra, with sincerity and full heartfulness, sound the mantra:

   ONG AIM HRIM SARASWATAYA NAMAHA 108 times.

   (Saraswataya is pronounced Sa ra swat ay ey, with a short "a" sound at the end.)

Apana: Release

Dual are the breaths vibrating here, returning to source and flowing afar; May one bring strength to you, the other blow disease away.
– Atharva Veda

Apana is our downward and outward flow of prana, part of the force of gravity. Apana is our exhaling, releasing and purifying, "blowing disease away" as the direct polarity to the inhaling, inward and upward movement of prana.

The archetypal polarity of Prana and Apana is continually expanding/contracting, opening/closing, pushing/pulling, bringing in/releasing. This is the flow of the universe: waves come into the body; waves exit the body. Our body-minds are always in a process of giving and releasing.

Apa is the "air that moves away" and disperses, like the exhaust of a car engine. Apana governs the exhalation and elimination of carbon dioxide through the breath, and releasing through sweating. Apana rules all excretory functions and is most associated with the sacral and root chakras, water and earth elements.

Apana is a release valve through which emotional, physical and mental waste can flow out of us. In this release of our baggage and negativity, we "let go" of toxic and repressed emotions, unneeded foods, physical toxins, mental habits, old attitudes, belief systems, and relationship debris that is no longer required for our growth and well-being.

This is not just a one-off movement: it is a constant process. We constantly need to let go of what is harmful and unhealthy for us. Whatever is no longer needed to sustain us and whatever is no longer useful to us for our growth and happiness have to be released. We need to constantly clear out our inbox on all levels to allow new information, new vibrations, fresh energies, new people and new ideas to enter our world.

Apana is perhaps the most vital pranic flow for our survival and healthy somatic, psychological and emotional functioning. In the Upanishads, there is a story about the five pranas in which each one tries to assert their supremacy. At the end of all four of the prana's arguments as to why they are the most important, Apana's time to speak comes. He looks up and says, "Without me you would all be dead, this body would be dead. I Am the Master of the body."

Without being able to release physically from your body, *you would die*: your body would turn grossly ill, and your body systems would start to shut down one by one. (Apana is part of the basis of our immune function.)

Without being able to release emotionally, one becomes depressed, frustrated and angry. Without being able to release mentally, one becomes fixated, rigid, isolated and trapped in your mind, becoming judgmental and fundamentalist. Without being able to release your spirit, you become identified and trapped in your body-mind, believing the world of the senses and the third dimension is the only reality.

To have an open and balanced flow of apana we have to investigate and release our façades, the masks of the false self that you project outwards to others in order to fit in and be accepted and to cover up your own emotions.

You cannot fully open apana unless you fully embrace and accept yourself, which means the façades and masks that you use have to be discovered and released. In this process, you come to deeply feel what wounds and emotions lie behind your mask, and begin to accept who you really are. Apana in its height is about true acceptance.

Apana points towards the most shameful, unloved, unworthy, embarrassed, humiliated, hidden and taboo parts of us. As its exit point is the anus, many perceive it as dirty, a place where we hide our judgments and sweep under the carpet our shames and fears, where they literally do not see the light of the day. Here is literally where our shit is; what we do not wish to see, what we do not want to admit and accept about ourselves.

Apana flows out of us through our anus. When we have a tight anus, we can be controlling, fearful, manipulative, emotionally closed, mentally rigid and judgmental in our attitude to life and other people. Excessive political and spiritual correctness arises from this area being closed.[170] Apana opens and flows more by exposing your fears, shames and judgments to yourself and others, by being vulnerable and honest. This helps you to let go of these emotions.

Letting go and ceasing to be "anally retentive", or contracted and closed, is letting go of all that which does not take you into happiness. Without releasing old buried debris and the past, we can never truly look upwards and forwards, and allow the fresh flow of Prana to enter and enliven us. We cannot move ahead with the new; we cannot live to our higher potentials.

Excretion, or letting go, is an archetypal quality seen in water, the element that is the governor of Apana. Water is always letting go and moving through that which is unmoving and stagnant. Water is infinitely adaptable, moving and flowing into any form. Letting go, like water, also implies transparency.

All physical wastes and toxins, emotions, thoughts, beliefs and ideas can flow through you, like water, when you do not hold onto anything. Little can irritate you or cause you to react negatively, as you hold onto nothing. If something does get to you, you can release it immediately and spontaneously in the moment, like a child does with their pains and hurts.

Blocked and poorly flowing apana reveals your own lack of awareness about yourself and the masks of your own dishonesty; it can hold a fear of reality. It holds where we are unloving to ourselves, and unaware of a need for change. When apana is unbalanced it can cause depression, clogging us up with experiences that weigh us down, making us fearful, suppressed and weak.

As Apana opens more, mucus, faeces, toxins and other physical substances may come out of your body. Hidden fears, shame and old humiliations may arise to be released. There may be a catharsis, a venting of ancient pent-up suppressed emotions, guttural sounds that need to be released, truths that need to be told, stories that need to surface, fears that need to be screamed out and exposed, vulnerabilities that need to surface behind the masks, family secrets that need to be exposed to the light of day.

As apana flows more, needed positive changes can occur as you are more honest with yourself, which in turn can lead to more self-love and satisfaction in this nurturing of yourself. We can then feel and express more clearly and do not so readily go into self-judgment, self-blame and self-doubt. We embrace our aversion and dislike of things within us, enabling us to embrace and accept the same qualities in others.

As apana flows freely, we accept more and judge less. We are more forgiving and open to everyone and their beliefs. Apana healing can allow the surfacing and transmutation of hidden sadness, sorrow and disease. The balanced flow of Apana can also protect us from negative astral and spiritual influences.

Apana Practice

When Apana is balanced and free flowing, our energy and sense of authentic truth can be expressed freely as it becomes *free of the rigid control of the mind*, moves spontaneously, and is unbridled by conventional restraints and cultural rigidities.

Mudras are specific gestures that connect into the meridians and organs of the body, stimulating their electromagnetic properties in order to awaken, vitalize, and clear these pathways.

Gently focus your awareness on your navel, anus, and the space between them. Inhale and exhale deeply into this space x9. Drop your whole attention into this space.

1. On both hands, bring your middle and ring fingers together on the tip of your thumbs.
2. *Focus on your navel-anus space and your breath, contracting and expanding the anus in time with your breaths.* On inhale, gently squeeze the muscles of your anus; on exhale, relax them. Start off slow and find a rhythm, then if you wish to accelerate the process, do fast breathing and fast contractions.
3. *As you inhale and squeeze,* remember a shameful, dirty, or guilty memory feeling in your life. Remember when you have had to stuff down your truth and suppress yourself. Remember when you have given away your power. Remember when you have been humiliated. Remember when you have been embarrassed or felt dirty and ashamed. *Remember it all.* **When you exhale and relax, focus on letting them go.** Keep repeating this rhythm and relax into it. Make it more and more powerful as you breathe it. Use your intention and desire.

To help you in this process, ask yourself these questions: Do you constantly criticize and self-punish yourself and others? What are you ashamed of? What is the most shameful thing you have ever done? Do you blame others? Do you feel unhappy with yourself and your life? Do you feel that no matter what you do, it is never enough? Do you believe you are not worthy of good things in your life? Do you feel betrayed that what you thought is true, no longer is?

Are you always a "nice" person, who represses what you really feel? Do you have fear about expressing yourself and what you really feel? Are there things that are too taboo for you to even contemplate doing? Are you living your life for others or for yourself? Do you have a lot of anger buried inside you?

4. Now, *with each inhalation and squeeze of the anus, sound the syllable of Apana: PH or PHA.* As you exhale out, release the old *and sound PH or PHA*, letting the sound current flow. Do for 5 minutes.

   To sound PH: Purse your lips as if to blow out a candle. PH sounds like "pharmacy".

5. STOP. Ask: *"Beloved Source/Father/Mother God, please help me release all my blame, criticism, anger, fear, shame, sadness and control out of me now."*

6. *Now ask these questions:* Do you feel guilt, regret, and uncomfortable with decisions you made in the past? Do you hold on to buried anger towards your-self and others? Do you fear loss and abandonment? Do you not forgive yourself or others? Do you stuff down your emotions, and pretend they are not important? Are you afraid of being the fool? Are you afraid of being wild and out of control?

   Do you hold yourself tightly and rigidly? Is your body stiff? Do you always follow rules and regulations? Do you do whatever you feel? Are you a control freak, controlling your reactions, your emotions and mind? Are you always concerned as to how polite you are to people and how good you look in front of people? Do you always try to fit into your social group or into society?

7. **Say:** "Beloved Source/Father/Mother God, please help me feel and release all my deep feelings of guilt, blaming myself for everything wrong that happens, punishing myself, and being angry with myself and others. Beloved Source/Mother/Father God, please help me feel and release all my fear that keeps me down and stuck in a box. Please help me feel and release my fear of power, my feelings of shame and ALL the deep humiliations I have had. Please help me feel and release my anger, my bitterness, my sadness, out of me now!"

8. Sound the mantra: **ONG HRIM KLIHM DURGAYAI NAMAHA** 108 times.

   (Durga yai is pronounced Durga yey, short "a" sound at the end.)

   Focus on your primordial brain at the back of your head opposite the chin, whilst doing this mantra internally.

9. Breathe in *SA*, filling yourself. Inhaling sustains life.
   At the end, rest gently and feel the holding of everything. Breathe out *HAA*, emptying, emptying. Exhaling purifies. At the end of this HAAA, hang in the balance, suspended. Touch the tip of your tongue to your palate at the roof of your mouth. Feel into what is directly above. Exhale *HAAAA* upwards. Savour this sound of perpetual creating.

Vyana: Circulate and Connect

Vyana is the pranic current that circulates throughout your entire body-mind. It is a connector, diffusing throughout your body-mind, communicating pranic information throughout your body-mind. *Vyana* means "outward moving", and includes the circulatory system, muscle and skeletal system, nervous system and nadi/meridian system. These systems interweave throughout all levels of the body-mind.

Vyana governs the movement of Prana through the Nadis, keeping them open, clear, clean and even in their functioning. Vyana is most connected to the manipura chakra, moving out from here to the peripheries of the body-mind. It also connects to the crown and auric field.

Vyana moves food, water and oxygen throughout the body, and in doing so it assists all the other Pranas in their work. Vyana permeates and extends, moving energies that have been absorbed and distributing them: it is usually expanding unless it is out of balance. Vyana connects all parts of your body-mind as it governs circulation on all levels: emotional flow, mental circulation or openness to new concepts, beliefs and ideas, as well as the physical circulation of specific nutrients.

Vyana allows us to keep moving so we do not get rigidly attached to anything, and can therefore be open to new inputs and experiences. Vyana flows everywhere, from the tips of your toes to the crown of your head and everywhere in between. Vyana helps create and maintain your etheric field, that which is closest to your skin just outside your physical body: the innermost part of your auric field.

In this sense, *Vyana helps you attract what is within you from the outside*, be that pleasurable or painful, as this information is within your body and field. Your auric field is nourished and kept circulating by Vyana, keeping it flowing and alive. If something is not moving within you, is stagnant, then this will be felt by others through your auric field, coming to them as an intuition or felt "sense" about you.

Indeed, most of us have gut instincts, hunches or senses about others *through our energy field sensing them first*, secondly by their body language and appearance, and then by their speech. This is because our auric fields extend out and away from our bodies, becoming the first aspect of ourselves that comes into contact with others. Keeping our circulation alive and well will mean others will feel us more clearly, and we can attract more of what we desire.

Vyana feels more like light within you than breath. The saying, "a rolling stone gathers no moss," means that in movement nothing can attach to us, harm us or stop us moving forwards. Vyana helps us by keeping energy inside and outside our bodies freely moving in spiralling toroidal flows.

When Vyana is unbalanced it may cause us to feel isolated and separated, disconnected with ourself and less able to connect with others. We may not feel connected to who we are, what we do, where we are and whom we are with. We feel something is missing if vyana nourishment is not flowing through all parts of us and not communicating information from one part of our body-mind to another part.

Vyana Sarvayatra: Pilgrimage Around the Body

Vyana is a connector between prana and apana. It circulates and moves subtle life-flow throughout the energy points of the body-mind. It can be accessed by mudra and sound.

1. Begin by sitting up, spine straight. Place your first two fingertips on top of your thumbs, leaving your little and ring finger straight up. This is Vyana Mudra.
2. Focus on your crown chakra. Sound *CH* or *CHA* for 3

minutes focusing here. Have your mouth open, tongue touching the top of the lower teeth. *CH* is pronounced like the English word "charm".

3. Staying in the mudra, follow your inhale to the third eye. Visualise a milky white colour as the breaths flow to the throat, the left shoulder joint, left elbow, wrist, tip of left thumb, index finger, middle finger, ring finger and little finger. Follow the circulation and inhale BACK through the wrist, elbow, shoulder joint, throat and crown.

Now: *Repeat for the right side of the body.* Staying in the mudra, follow your inhale to the third eye. Visualise a milky white colour as the breath flows to the throat, the right shoulder joint, right elbow, right wrist, tip of right thumb, index finger, middle finger, ring finger and little finger. Follow the circulation and inhale BACK through the wrist, elbow, shoulder joint, throat and crown.

Now breathe down to the heart centre, feeling the flow between left nipple and heart, right nipple and heart. Travel down to the belly button and focus on the base of the spine. From here, breathe through the left hip joint, down the thigh to the left knee, down through the ankle, big toe, second toe, middle toe, fourth toe and little toe. Inhale back through the ankle, knee, left hip joint to your spinal base. Relax here for a moment.

Now, repeat for the right side of the body. Breathe down to the heart centre, feeling the flow between right nipple and heart, left nipple and heart. Travel down to the belly button and focus on the base of the spine. From here, breathe through the right hip joint, down the right thigh to the right knee, down through the right ankle, big toe, second toe, middle toe, fourth toe and little toe. Inhale back through the right ankle, knee, right hip joint to your spinal base. Relax here for a moment.

4. Follow the stream of Vyana up from the spinal base to your navel, heart, throat, ending at the pineal and crown.

5. Sound the mantra: **ONG AIM HRIM SARASWATAYAI NAMAHA** x108.

   Pronounce Saraswatayai Sa-ra-swa-ta-yey, short "a" sound at the end.

Udana: Ascending Prana

The control of Udana is the control of all five pranas... and leads to the Supreme Self.
– SD 2:568

Udana is the Ascending upward directed prana flow that rises from the throat chakra into the whole brain. It rises upwards, directing its flow into the glands of the brain. Udana Vayu means "upward moving air", and is the most spiritually elevating of the five pranas. In the Prasna Upanishad, it is *Udana* which guides the mind from the waking state, into sleep, into deep dreamless sleep, and beyond into unity consciousness.

When Udana is balanced it becomes a luminosity and presence, a sublime flow. As it rises and stabilizes, one is no longer entangled in the physical world. Udana is ruled by air and space.

Udana empowers the throat chakra. It is the air that rises ever upwards, and does so from conception through to, and beyond, death. At the death of the body it moves the spirit out of the body to the astral and causal planes. One can do this consciously whilst still having a body in astral and soul travel.

Udana is the most important Prana for sound vibration. Udana empowers speech and is the prana that can refine our speech. Advanced meditators use Udana to tune into and merge with universal currents of sound through inner mantra in the mode of sound *Pasyanti*. When we make sound from this super-conscious state of Pasyanti, one can enter the void *and* can emanate sound outwards from this space. This sound deeply affects all those around you, as well as propelling the singer into the void.

When Udana is out of balance it can result in negative, harsh, bumbling, crass, inappropriate or excessive chatter.

Conversely, it can also result in not being able to express yourself or speak clearly. Unbalanced Udana may hamper the intake of physical nourishment and cause pride and arrogance. We may become ungrounded, trying to become too "spiritual" and losing track of our roots.[171] We can become too heady.

Opening the flow of Udana can help you express yourself more, express your truth and express what you have blocked or shut down from expressing if you desire this. When Udana is balanced and strong, we stand tall and are enthusiastic, alert, articulate and strong-willed. Udana helps awaken spiritual and creative potentials.

Udana affects the mind and brain, bringing clarity and lightness. In Jnana Yoga, the path of wisdom and knowing, one needs strong Udana. Without strong Udana you cannot succeed on this path, as it involves Self-Inquiry, the process of inquiring, questioning and witnessing what is happening within you and in the world.

True Self-Inquiry is not just a mental exercise, *but also an energetic flow.* You need a good flow of Udana in order to be able to question, witness and inquire. This is why many people cannot inquire deeply into themselves or question the world around them, as their Udana is poorly flowing and stagnant. The mind is fuelled by prana, and to become more conscious entails sustained self-inquiry which is fuelled by clear flowing Udana.

The first time we consciously experience a surge of Udana may leave us lightheaded, expanded, feeling headless, dizzy and ungrounded. This is a good sign at first as it means the Udana is moving out of a state of poor flow. If these symptoms continue, it is time to ground yourself with the other pranas.

Udana Practice

1. Sit up, spine straight, in Udana Mudra on both hands. All four fingers except the little finger are touching the tip of the thumb. Focus on the throat chakra.

2. Breathe *Ujjaiyi breath*, a rasping breath that moves against your throat. As you do, breathe up from navel to heart, and as it reaches your throat, tilt your head back and up. Look towards the heavens! Draw this energy up into the pineal. Contract your anus and hold the breath for 3.

3. Exhale down from the pineal into your throat, heart and navel. As you reach the navel, do *jalandhara bandha*: draw the chin in and down, gently compressing the front of the throat. Contract your navel and anus inwards. Hold the breath for 3. *Repeat this ascending descending cycle for 5 minutes.*

4. **Sound the seed syllable for Udana, TH,** into the throat 36 times aloud. Then: Sound *TH* into the third eye for 5 minutes internally. *To Sound TH:* Your mouth is wide open, the tip of your tongue touching the cutting edge of the upper teeth. In English, it is pronounced like the word "faith".

5. Sound the mantra: **AUM AIM HLIM MA** in the throat x108.

   AUM AIM HLIM AM in the pineal 108 times.

Samana: Fire and Digesting

Samana is the fire of centred Vitality, digesting, absorbing and assimilating energy and information. Samana fire converts everything coming into you into appropriate energies. It also separates energies in order to bring them together in appropriate ways, and acts as a translator to bring energies into balance.

Samana's heartbeat is at the manipura solar plexus chakra: the hub of energy, vitality and the transforming power of fire, and includes the stomach. Samana maintains the vital heat of the body and is the gastric fire. It regulates your inner fire with fuel, which must burn evenly.

It assimilates Prana like a power station, digesting energy absorbed through breath, food, the five senses, emotions and mental experiences for *Samana is digestion on all levels.* "The fire that is within a man and digests the food eaten is Vaisvanara."[172] This fire digests food in the gut, air in the lungs and works with the mind to digest experiences.

Samana moves from the periphery to the centre through a churning and discerning action. It mediates between what comes in, *prana*, and what goes out, *apana*. It extracts nutrients while leaving toxins behind. When we have good Samana we have a healthy, vital glow and radiance exudes from us, with bright eyes.

Samana is a concentrating, absorbing and consolidating force. Samana fire centres us into a quiet, balanced and peaceful state when used properly. It nourishes, bringing contentment and balance. When we sit by an open fire, we can feel calm, soft and at peace. We can do this internally too, in Samana Yoga.

Without the balance of Samana we cannot sit in our being, come into peace, or concentrate the mind. Deeper meditation is not possible without healthy Samana, the "balancing air", which is why yogis focus on this pranic flow in tapasya, or deep meditational practices.

Samana is the fire of reason and discernment, sorting, digesting, discarding and arranging, and when harnessed it can burn away impurities. Digesting is a form of discernment, so *Samana* is a continual and ongoing process of being discriminating in our choices and life. Samana gives us clarity and courage to see ourselves. Our power of choice is fuelled, guided and actualized through our pranas and mind, supported by the inner, centring and vitalizing fire of Samana.[173]

When Samana is healthy, we benefit from strong digestion, vitality and balance at every level. When Samana is weak, we struggle with assimilating and digesting what we take in and what we consume in our physical, emotional and mental experiences. We cannot process our emotions, we cannot process, digest and learn from our life experiences, we cannot digest our food properly to our benefit. We cannot be nourished.

When Samana is unbalanced it deepens the sense of attachment and greed, making one *consume more*. We cling to things and become possessive, more materialistic, more outward focused, more busy.

Samana can also help connect one into the universal creative fire of Agni, with the highest form of Samana residing within your heart as a flame. As we develop this energy further, it supports the creation of more *Ojas*, vital energy, *amrita* or essence energies, and *soma* – the elixir of bliss. These are three means by which Unified Consciousness can manifest in the physical body.[174]

Samana Practice

As you hold your breath in the space between inhale and exhale is where the embers of samana get fanned into a flame. The following sequence purifies samana, fanning these flames.

1. Stand with your feet 6 inches apart. Bend slightly forwards, resting hands on knees. Become aware of your breathing and relax. Exhale and contract your belly muscles, drawing them in and up. As you inhale, relax the muscles. Do x20.

2. Bring the right thumb and ring finger to close the nostrils. Take two forceful exhalations and inhalations (bhastrika) on one side, switch, and immediately take two forceful breaths on the other side. Alternate for 2 minutes.

3. Do Uddiyana Bandha, pulling the abdomen up and in, bringing apana toward the solar plexus and lifting it into prana. Do this x6.

4. Sit up straight in Samana Mudra, where all your fingers are touching the top of your thumb. Lower your head slightly and focus on your stomach. Sound the **seed sound for Samana, TH,** for 3 minutes. To sound TH: Open your mouth with your tongue touching the top of the upper palate.

5. Sound the mantra: **ONG KRIM KLING KALIYAI** 108 times. Physically contract the belly button area when you make the sound **K** into the stomach.

The Trinity of Mind – CA TA TAU

The causal seed of the mind, intellect and ego.

The trinity of the mind is fuelled and moulded by the five pranic flows. This trinity is composed of *manas*, the bridge between physical reality and our mental data processes; the *Buddhic* mind of intellect, discernment, choice making and will, and the ego-identity of the separate self *ahamkar*.

All three interconnect, and are fuelled and powered by the five pranic flows in a circuitry that connects our body-mind as one system. All five prana flows are symbiotic and evolve as one process with the Trinity of Mind, which stabilizes the pranas. The body cannot live without the mind "for mind and prana are twins in function". In this context, the body's energy flows need the mind. Without these bodily energy flows there is no mind, without mind there are no bodily energy flows. If one dies, the other dies.

The mind is intimately connected to the five pranic flows and the five elements: it can be spacious, it can flit like the air, be energetic like fire, flow like water and be stable like earth. A clear efficient mind arises from the balance of all these forces.[175]

Mind needs the pranas as its energy source and distributor for the five elemental energies, and together, mind and pranas create a "wireless circuitry that interconnects every point of the subtle form with every other point in the energetic system."[176] With imbalances in the life-force, imbalances in mind occur: with imbalances in the mind, imbalances in life-force occur. Both are symbiotic.

The Trinity of Mind stabilizes our individual identities, giving us the capacity to say yes and no and make choices. Our freewill is actualized here, manifested by the vital flows of prana. The mind shines with the reflected light of its source consciousness, the Sun, thus the mind is likened to the moon. The mind is not the Cogniser, just a necessary instrument of cognition. If we only identify with this reflection, we miss the source of Awareness, the One.

Ca: Manas – Mind

Manas receives data from the physical world, importing sensory impressions into the brain, and then reconfigures this data to project it outside to form a picture of the world. Manas is the intermediary between the physical world and your brain.

Manas identifies objects and names them through thoughts. "This is a wall, this is a tree." It is only through this mental contact that one can understand objects.[177] Manas is the faculty that cognises the objects perceived by the five senses and speech. When there is no contact of manas mind with the sense organs and their objects, no grasping, comprehending or understanding of objects can occur. This also means knowing the difference between inside and outside.

If mind is not connected to the senses, the senses cannot make any sense! Looking at or sensing something means nothing until *manas* does something with that data. Manas organizes thoughts and functions to navigate the 3D world, and is the structure that keeps our 3D life in order. It creates lists of things to do and errands to run.

Manas forms mental boundaries, which are necessary to allow any form to be perceived and worked with, yet when this aspect is overactive it generates a sense of confined limitation because the sensory world becomes your primary reality.

Manas cannot make decisions or implement anything without being instructed. It brings raw data into knowledge and needs the discerning intellect of Buddhi to make informed choices. If we are not connected consciously to Buddhi, then manas will get its instructions and guidance from somewhere else, like the external world, other people, the media, your subconscious needs and more.

Manas is object orientated: it considers the bodily perspective on the pros and cons of a matter. In language, it is the noun, that which objectifies things, making them solid and concrete in the 3D. When this autonomic mind is in its right place as a servant of the Buddhic or mindful mind, it performs its rote and necessary functions admirably. When manas rules, gossip, mental chatter, babble, mental disturbances, materialism, narrowmindedness and a sheep mentality occur.

Overreliance on manas dominates modern day society, generating fear of change, fundamentalism, short-term thinking, rigidity, adherence to dogma and aversion to anything "out of the box". Much of modern media, culture and science is based on manas rather than Buddhic intelligence. Relying on manas without Buddhi makes us easily controlled and controllable as manas only considers what is known, your own personal accumulation of knowledge and what you have learnt already in and from the world.

Unbalanced manas is the automatic, babbling mind that goes on and on; all sorts of banal thoughts arise and keep coming. It may seem like these thoughts are annoying and time consuming, but it is the resistance to these thoughts that cause discomfort or irritation. The problem is not the thoughts but the effort expended in trying to stop them.

Manas mind, in harness with the egoic I *ahamkar*, will always be trying to do something, but when manas is "freed from the egoic ahamkar, it can turn from the objects of the physical world inwards toward its primordial ground, which then leads it to enlightenment".[178]

Awareness never changes; it is the interpreter of manas where change seems to take place. Manas is always in a state of flux. In this state, it cannot receive pure consciousness. Continuous emptying of our mental inbox and alignment to the Buddhic mind and beyond is necessary to receive new and higher knowledge and wisdom.

Sounding Manas: CA

This is a palatal sound with no extra breath. With mouth wide open and tongue raised to slightly touch the lower teeth, say "cha".

Ta: Buddhi – Intellect and Discernment

Self rides in the chariot of the body; Buddhi the sure-footed charioteer, manas the reins.
– Katha Upanishad 1:3, v. 3-4

Buddhi is the mind that makes something known, the guiding force of the trinity of the mind that enables us to perceive, understand, speak, discern, choose and act beyond manas and ego. Buddhi is learned knowledge; in its highest aspect, it applies the knowledge and principles it has learned through intelligence. Intelligence **is** the ability to apply any knowledge in the present moment in a spontaneous way through the process of *Information, Knowledge, Wisdom and Knowing.*

Information is inputted into your system from reading, listening, media, Internet, TV and receiving data from others in a sensory based viewpoint of manas. *Knowledge* takes this external information into you so you can begin to understand it. You are absorbing it, making deductions, assertions and analysing according to your own viewpoint, or ahamkar – the egoic separated self. With a degree of detachment, witnessing and overview of both of these aspects, we access the Buddhic intellect.

When this information and knowledge are brought into the Buddhic intellect and is synthesized with direct experience, we arrive in wisdom and *Intelligence* – flowing, spontaneous and in the present moment. *Knowing* or *Chit* arises when your mind is empty of contents, and is directly connected to the Universal Prana.

Buddhi in its highest aspect means becoming awake, awakening and being open. The word Buddhi arises from *Budh*, to awaken and recover consciousness, hence Buddha: the enlightened one, the one who has awakened. For example, when you experience an insight – an "aha moment" – you have a moment of transformational understanding. Maybe it will last a second, a minute, a month, or a lifetime. Yet for that space of time, you recognise truth as a direct experience and it changes you.

But: later on, your manas and ahamkar ego mind may become involved, and your immediate experience of direct Knowing becomes filtered into an abstract idea, concept, or a new "rule" to live by. The mind may further reduce and catalogue the experience down to a bit of data, an abstract intellectual piece of information.

The ahamkar or ego mind's assumption that it "knows" or believes a truth is separate from an actual experience of it. In the true Buddhi, you actually *live* what arises in any given moment, because you are present with what Is. The ego mind of ahamkar is a servant of the buddhi when it is in its rightful place, and then it is a good tool. Always keep it in this place. This is wisdom.

With Buddhi as the master of the Trinity of Mind, you direct and use the ego-separate self of ahamkar and manas mind to structure or plan things, thinking with your heart. You think how to put the heart into a plan of action and how to apply love in your life. The ego mind ahamkar becomes an autonomic tool for buddhi to express, organize, structure and manifest the soul's vision, the soul's purpose, the soul's dharma and passion on Earth.

Ahamkar as a servant of the buddhi and soul means it is there to do the bidding of the soul. When this is integrated, you will not have to think about 3D things much of the time – you will just do them automatically. You will be able to answer emails, pay bills and do mental tasks without thinking or getting caught up in them. You act precisely, efficiently and quickly when the ahamkar mind is the servant because Buddhic mind is focused in a relaxed, aware way, not in an intense, rigid way.

Buddhic mind considers and eloquently articulates all aspects and facets of a problem, situation, subject or theme, just as an expert can take one paragraph to explain succinctly what others would take 10 pages of writing to do. Buddhi is always willing to consider evidence and truths that contradict what it knows because it is neutral, open and humble.

Buddhic mind can assimilate and understand any body of knowledge. It can follow the threads and see the depth and multidimensional facets of any subject. In quantum physicist Amit Goswami's understanding, there are three layers in the Buddhic mind:[179]

1. *The creative, psychics and mystics.* These people are outwardly creative, musical, artistic and healers. They regularly experience the paranormal and metaphysical and make evolutionary leaps in their lives. The shift of identity from egoic mind to Buddhic mind has started but the primary focus is still on the external world.
2. *The Transpersonal.* These people are focused on giving to others. Social, cultural and psychological aspects no longer have real meaning to them. One honours others as being unique, different and valuable, seeing that we all come from the same consciousness. Stabilization in Buddhi is happening.
3. *The Buddhic.* One surrenders the idea of serving the world or others. Individuality is seen as a necessary function but not innate. Action lies in the moment and has no formula to it. One is whole within Self, with no need to be anyone or do anything, allowing whatever is meant to happen to happen.

Humility becomes integrated. Inner creativity reigns with outer creativity no longer having any charged meaning. Seamless integration of all parts of mind occurs without conflict. One's identity, way of thinking, and way of being by oneself becomes fluid, without egoic concern. Surrender occurs.

This leads into *Cit*, the fourth aspect of Knowing, beyond memory: freedom from ANY mind, and the Awakened Self. This intelligence of pure awareness is beyond divisions, karma and who you believe yourself to be. In Cit, you live life in the moment, fresh, pure, innocent and immaculate. Cit is the aliveness of awareness, beyond the mind, intellect or ego self, yet using this Trinity of Mind and the five pranas as servants.

Discerning Mind

The enlightened Buddhi knows when to engage and when to withdraw from action, what should and what should not be done, what is to be feared and what is not to be feared, what actions lead to bondage and which lead to liberation.[180]

Buddhic discernment sees the whole picture of a situation because it has no judgment or emotional charge. The difference between judgment and discernment is the assignment of value – the value of being right or wrong, good or bad. Judgment divides and labels. If you have an emotional charge or wound, then judgment occurs and this is part of ahamkara or ego mind. Buddhi witnesses what is occurring from a neutral place.

Discerning wisdom is calm clarity, and it has no charge to it. This is quite different to judgment. Judgment is the beating heart of duality. Our judgments are a hidden blessing for us to see, as they show what parts of us are still mired in emotional wounds, traumas and memories that create the sense of ahamkar.

In judgment, you have an emotional sting and hidden pain that is projected outwards. You compare yourself to others: who is better, who is worse, who is right and who is wrong. We step out of Buddhi the moment we judge, attack and defend. The truth is, you benefit most from not judging, as it leads to the peace and freedom of Buddhi.

To be discerning we have to be self-responsible and take hold of the reins of our reactivity. We have to be open to what we are feeling and viewing without judgment. In discernment, we fully own the emotional sting in how we feel and view others, and we see how we project parts of our unseen and unrecognised self onto others. If you are faced with anger, you can choose to respond with peace instead of react; if you are faced with your compulsive unconscious actions, you can choose to not follow them and chart a different course. This is Buddhi.

Buddhi sees all events and people as neutral, is openminded about what is happening, and faintly curious. It chooses to not judge others, realizing we make people and situations good or bad through our own egoic ahamkar filters.

Choice

Discernment distinguishes the truth and effects of a matter, person, timeline and choice of action. It sees the effects of potential choices you can make *before* you put them into action, and can ascertain what these choices and course of action will lead to. This then allows you to take charge of your life by making informed conscious choices.

Buddhic discernment can create courses of action to achieve what your soul wants, and what you practically need to do to achieve your goals. It decides what is necessary for the soul to do in order to reach its objective. What is necessary for me to do today in order to achieve my objective of being Self-Realized?

How clearly we are in touch with Buddhi is the degree to which conscious choices are made, and conscious actions generated from these choices. One performs an action only after making a decision, which comes from Buddhi. Buddhi knows the right time to do or say something, as well as the right time NOT to say or do anything.

Buddhi in its pure state is a sharp, clear, open mind. It is selfless and intuitively understands which actions lead to more love and freedom, and which actions do not. The frontal lobes of the brain are associated with the buddhic mind. When the frontal lobes are fully activated it is the highest form of mental function we have. On a subjective level, the signs for underactive frontal lobes are dullness, weakened will, a follower mentality, submissiveness and boredom.

In order to rise to a higher order of consciousness in your life, you have to dissolve the old order. In evolution, you reach a new plane of order and harmony in yourself for a while, and then you feel a need to dissolve this order and move to the next higher plane. This process continues throughout your life as you evolve. No problem can ever be solved at the same level of the problem. One has to go to the next octave beyond the problem in order to resolve it by seeing clearly what is possible and what is not. This is Buddhi.

Discernment allows one to contemplate things, events, situations to see if they make logical sense *as well as* sense from an awakened perspective, as true discernment uses both. Discernment is also about value. What is worth your precious time and energy to invest in?

Discernment is knowing yourself and your capacities and what is do-able and achievable. Discernment allows us to nurture and harvest our energies wisely. Saving time, unnecessary heartache and life-force energy is a valuable aspect of discernment, ensuring you do not keep repeating the same patterns and mistakes over and over again.

Buddhic mind is spacious and fruitful: it doesn't chase dead ends. It allows you to let go of things that can never work, and persist with those things that can work. Discernment saves us pain, heartache, loss and time.

Discernment brings efficiency and economy of action and effort to our lives. It is the ability to detach from, and have a clear overview of our life, relatings, soul purpose and choices. Buddhi helps us to grow, to become aware, to be mindful, to make informed and conscious choices and navigate our lives as conscious, aware beings. Without Buddhi, we remain in the ego I of ahamkar, run by our own unconscious tendencies, patterns of memory, beliefs, traumas and conditioning.[181]

Buddhic action is a considered response (not a reaction) situated in a non-judgmental, compassionate outlook. Once consciousness arises with a discerning mind, it is a powerful possibility: this is what makes human life unique and capable of enlightenment.[182]

Buddhi makes *conscious choices and can enact that which leads to freedom*. Buddhi investigates, contemplates, takes new directions and comprehends new discoveries. Buddhi can discern the true nature of anything. "True discernment is the capacity to distinguish between the bound and the boundless, between the psychological and the existential, between delusion and truth, between your relative perception of life (Maya), and life itself."[183]

Discrimination of what is real and unreal involves a transformation not only of thought, but also of emotions, feelings and soul. That is why we find, amongst the great beings that have awakened, a total unity of thought, word and action expressed through the refined Buddhic mind.

To sound Buddhi TA

TA is a cerebral sound. Focus on both of the frontal lobes of your brain. Your tongue tip touches the roof of your upper palate with *T*, and then flicks forward afterwards, your mouth opening in *AH*. *TA!*

Meditation

Imagine you are an eagle with clear sight, soaring above your body, soaring above the room you are in, soaring above your house, soaring above your city, soaring above your country, soaring above the earth. You can clearly see the web of people, relationships, work and home you live amongst, *but now you are not part of it.*

Let the eyes of Buddhi give you a fresh perspective on it all.

Tau: Ahamkar – Ego – Separate Identity

Self is the Aham, but ahamkara is qualifications like body and name. In ignorance, one is possessed by such qualifications and superimposes them over the Self as if one is possessed by a ghost, and thus behaves in the spirit of the ghost.[184]

Ahamkar causes you to think, "This is mine and this is not mine." *Aham-kar* literally means the "i-doer", the actions that come from your sense of i, the emotionally conditioned, culturally consensual, ancestrally inherited, memory-bound self that believes that it is all that it has experienced, and stays locked into this filter bubble or cage of experience. The I believes this to be reality, and moves through life navigating in this way, without discernment or questioning.

Ahamkar is a series of moving activities, thoughts, feelings, beliefs and memories bound together by the moviemaking narrative power of Maya. Ahamkar is not a solid entity – it is a flow of experiences and is formed by experiences. Aham-kar is the vehicle for all the effects and consequences that come from these experiences and our self-centred actions, also known as karma.

Most people's experience of their self is so mixed up with the content of their experience that they do not experience their Self clearly. This mixture results in a conditioned self that is qualified by the limitations of experience: the ego or separate self. It is on behalf of the ego that most of our thoughts and feelings arise, and in the service of whom most people's activities and relationships are engaged.[185]

The parts of the brain associated with this ego are the parietal lobes, which house the OAA, the Orientation Association Area.

The function of the OAA is to give us orientation in space. You may take it for granted that you can tie a shoelace or walk through a door, but this is only possible due to the neurological activity in the rear part of the parietal lobes. Brain damage to this area makes the smallest tasks like picking up a glass of water impossible, because the brain cannot perceive a distinction between the hand, the glass, and the space in between.

We need our ego ahamkar to perceive boundaries and distinction: to feel our individual body and to experience time. We need to feel separate and grounded in our own self-identity in order to be able to function in the world, to walk, talk, speak, move, relate to others, and to do worldly activities.

However, in most humans the OAA is chronically overactive. This stimulates the amygdala-hippocampus connection, a pair of brain centres designed to give a sense of meaning to perceptions and events registered as important. If the OAA, which is designed to create a perception of distinction to a useful degree, is hyperactive, we interpret this hyperactivity by assuming that *separation is the only reality*.

With a hyperactive parietal lobe, we perceive that we are fundamentally and existentially separate from everything we perceive, be it hand, glass, person, earth or universe. Ahamkar is the sense of self created by this neurological function, which is constantly in reaction to what is perceived as other than self, i.e. everything and everyone else.

Neuroscience has shown that in meditation or prayer the OAA is temporarily blocked from neurological input. This brings one into states of peace and expanded consciousness beyond any sense of self. People who have a non-functioning OAA *all the time* are unable to function in the world.

They are drunk in bliss, and unable to do basic things. Everyday functions like going to the bathroom become hilarious and nonsensical events as this consciousness has no body and no separate self. Being able to talk becomes a major achievement. Such people need help from others to even walk around properly! I have witnessed this in yogis in India, as well as in myself for two months every day when in the state of samadhi.

Ahamkar is the biological "former" of every cell in your body. It is your body-mind's organ of action, and without it there is no action you can do in the world. It also becomes your sense of healthy psychological boundaries, self-esteem and self-respect.

In a beautiful analogy from Peter Harrison, he shares that when one calls a taxi, one is in fact calling the driver to whom the taxi belongs. The taxi vehicle without the owner could never respond, and can move only when the being that steers the wheel decides to do so. All such calls come under ahamkar.

In Patanjali's Yoga Sutras 2.6, it is said, *"Egoism arises from the apparent identification of the Seer with the energy of seeing."* The seen world is in a continual state of flux, always changing. When we identify with this sensory mind, we enter desire and aversion, the twin poles of ahamkar: the resistance to change and the survival of the ego at all costs.

From *Aham*, the I Am soul or Seer, comes *ahamkar*.[186] Without Aham there is no ahamkar. Without ahamkar there would be no sense of the body-mind and no way of perceiving or navigating with the body-mind in the world.

Ahamkar perceives the changing universe of space-time and its qualities of flux, bringing individuality from the sea of space, bringing this into Self-expression. It is necessary and useful if ahamkar is the slave to aham, and aham is the master of ahamkar. The Seer of Aham is in the right place as the witness of ahamkar, and can thus guide your life in an evolutionary way.

One trick ahamkar uses to continue its survival is to create a fear based duality mindset and project its own issues and fears "out" onto others, blaming other people, the outside world, relationships, governments, partners, and even God for its problems, making others responsible for its suffering.

As Jung shares, "a man... is fooled by all the illusions that arise when he sees everything that he is not conscious of in himself coming to meet him from outside as projections upon his neighbour." Through this projection ahamkar only hears and understands others through the filters of its own thoughts, experiences, beliefs and conditionings, and seeks out other ahamkars who resonate with its view, further reinforcing its own sense of separation.

Ahamkar is partly composed of our memories. The many layers of our memories obscure our real nature as we identify with each picture-feeling-imprint experience of our past as being who we are. Each single, individual picture-feeling-imprint experience or photo of your life is woven into a moving narrative by ahamkar, *creating a moving picture or movie* of our lives. This is all thanks to the moviemaking power of Maya, the director of the movie, with ahamkar as the superstar actor.

This narrative of our lives is built within our minds in a series of snapshots given motion and coherency by ahamkara and Maya in order to create a movie of who we believe, think and identify ourselves to be. This could also be called our programming.

Memories individuate us and make us unique. They make us who we are today, they make the ahamkar. Memories can also imprison us in the past and create a narrative of life that is self-defeating and limiting. Memories create definitions and limitations of ourselves, others and the world, and whilst they can work for good and for bad, they do influence our lives and relationships.

The seed of memory was first conceived of in *RLRK* by Maya as one of the five vrittis, or mental movements, which help form the sense of the individuating I Am, and now the i of the ahamkar. According to Sadhguru there are eight types of memory, which I have added to:

1. *Elemental memories* recorded in the five elements of space, air, water, fire and earth within us, and how they are configured. We can change and erase these memory imprints by meditative practices and therapies, as well as by using the elemental mantras of the Maheshwara Sutra with a clear intention or sankalpa.

2. *Atomic memories* recorded in your atomic structures. "The dance of the atom in each individual is distinct because of past memory."

3. *Evolutionary memories* composed of the history of the human race and our planet, memories formed by all the collective actions and conditionings of humanity throughout history. From the dinosaurs to the Internet age, everything that has ever happened here on earth is recorded within our own fields in interconnection with earth's electromagnetic fields.

   Did you ever wonder why children enjoy playing with dinosaurs, even though they have been extinct for tens of millions of years? They were the original inhabitants of the planet, and this is recorded in our evolutionary memory. Our evolutionary memory has shaped our biology.

4. *Genetic memories* inherited from your parents, grandparents and ancestors. This culminates in the Rishi who was the progenitor and head of your bloodline, the original creator of your bloodline. Their memory also lies within you.

5. *Karmic memories* – present and previous life actions and their samskaras, seed impressions and patterns left in you from these actions, subconsciously influence your present-day actions, interactions, relatings and tendencies.

6. *Sensory memories* – what your senses enjoy become imprinted within you, and guide you to fulfil more of that sensory pleasure. The sensory overload in today's world influences how our bodies, minds and pranic flows

operate within ourself and how we react to others and the world.

Another way in which memory stores within us is through our skeletons, which are porous and have minute sonic chambers within them where sounds resonate, echo, and are "cached" by the body.

Our voices, tones and words carry electromagnetic charge and resonance. If you voice consistent negativity, and spend a lot of time in negatively emotionally charged environments and around negative people, these minute sonic chambers in the skeleton store this energy as memory through the process of resonant entropy. These emotions and sounds also resonate with the corresponding frequencies stored in the earth's fields.

Similarly, living in city "electro smog" 5G and constant Wi-Fi affect our ability to rejuvenate. The "quality" of our bones' resonant memories (bone RAM) and the electromagnetic charge of either harmonic or disharmonic frequencies affect us by being sonically "remembered", and stored in our bones.

7. *Inarticulate memories*, such as your emotional life in the womb and other deep emotions you have had that you cannot articulate, as they are preverbal, preconscious and prenatal.

8. *Articulate memories* – what you have experienced and expressed in this life that has consciously guided your beliefs, your life's course, and direction.

To Sound Ahamkar TAU

The tongue touches the bottom of the upper teeth in a dental sound with minimal breath. The *T* sound is slightly more pronounced, with the mouth opening with the *A* sound, to then close into pursed lips (as in blowing out a candle) with the *U* sound. One can pronounce this as *TAU or TAV*.

The Pranic Mind

"Vargadvitīyavarṇōtthāḥ prāṇādyāḥ pañcha vāyavaḥ
madhyavargatrayājjātā antaḥkaraṇavṛttayaḥ."
"The five winds beginning with prāṇā, arise out of the semi-vowel
in each group (h). The consonants are prāṇā and the five winds.
The subtle moving thought flows of the inner perceiving organ
antahkarana emerge from the first syllables of the three interim
Groups." (ca, ṭa, ta)

The consonants are *prana* and the five flows of prana: *K, P, C, T,*
T. The sound connecting all five pranas is *H,* the first of the semi
vowels *and* the sound for Space element HA. It is from Space
that the pranic flows arise, and *H* is the bridge between the
vowels of light and the form making power of the consonants.
Thus, we have *K* becoming *KH, P* and *PH, C* and *CH, T* and *TH,*
T and *TH.*

The *antahkarana* is the inner substance or medium in which
the movements of the *manas* mind, the *Buddhic* mind, and the
ahamkar mind flow. The subtle movements and flows of the
antahkarana evolve from these three functions and consonants
of *C, T, T.* The Trinity of Mind and Five Pranas fluidly and
harmoniously communicate through the antahkarana.

The Pranic Mind is the natural, organic framework for the
cohesive continuum of creation to flow through you. The Five
Pranas and Trinity of Mind combine in order to harmonize and
regulate the nervous system, moving one out of disorganized,
scattered attachments and external validations into internal
stability, internal reference points and inner sovereignty.

In this state, one resources more from within oneself and
less from others and the world. You acquire more awareness of
choice and more freedom of choice, and become proficient in
navigating life from your own life-force and clear mind.

When this Pranic Mind is operating optimally you do not get stuck in thought patterns, dis-ease, stress, anxiety or become dominated by any of these three minds: they all flow smoothly as they have been designed to, performing their functions through the antahkarana. The Pranic Mind arises from space element and connects your body-mind with space. Space is the medium for physical creation, and is everywhere in creation as an extension of the quantum code established in *AIUn*.

This means the mind will not dominate your life, allowing your antahkarana and Buddhic mind to flow, with the ego as its servant and the mundane mind performing its tasks automatically. Everything is in its right order and harmony as you are in flow with space, the medium and interconnecting substance of all creation. This then allows *Cit*, the fluid intelligence of awareness *AIUn*, to flow, resulting in you not identifying as being the Pranic Mind, but rather the awareness or oneness in which it all arises.

The Toroidal Pranic Mind

It is hard to surrender to a greater force than our separate self *ahamkar*. One needs to be in the flow of the toroidal pranic mind that connects to the universal toroidal pranic flow in order to let go of being ruled by ahamkar, and surrender to the awareness behind it.

We can understand this biologically and psychosomatically when we look deeper at the Pranic Mind. All the body's fluids form a unified wave-field. Through this fluid body flow tides, which mould and organize the physical body from the creation of the first cell.

The "Long Tide" within the fluid body allows life to take form,[187] and is the medium of our *primary respiration*, situated in the centreline of the body.[188] This Long Tide then extends outwards into space in a *toroidal wave-field*, which then connects into larger toroidal wave-fields streaming through space, to eventually connect with the universal toroidal flow begun in *AIAUch*. The whole pranic mind system is suspended within these spirals within spirals of toroidal waves, that spiral out further and further from our bodies into the origins of our physical universe in the "big bang" of *AIAUch*.

The Long Tide is a vital presence within and around the human body. Manifesting through toroidal wind-like spiralling forces, it generates local ordering fields that mediate embryonic development, igniting ordering powers in the fluids of the body and maintaining cohesive balance throughout life. It moves from the outside in, and renowned somatic expert Peter Levine calls it a *coherency wave*, as it maintains the harmonic coherency of the human system in the midst of all conditions.[189]

The Long Tide manifests *"like a great wind arising within a vast field of action, radiating through everything."* The Tibetans call it the *Unconditioned Winds of the Vital Forces*. It is unconditioned as it is not affected or conditioned by our personal experiences or conditioning; it maintains the pattern of the original toroidal wave or Breath of Life (generated from *AIAUch*) *no matter what conditions are present.*

We experience this Presence as a stream of light, a sense of awe, stillness, light and spaciousness, allied with a sense of interconnection with all life. Buddhists call this the mutuality or co-arising nature of all things. Consciously tuning into this Presence undoes much of our body-mind and somatic conditioning (which manifests through our pranic mind) as it brings us into resonance with the original toroidal wave-field.

Austrian scientist Viktor Schauberger recognised the Long Tide as *the Original Motion of Creation*, the movement of the One in action, working through the fluids of the body and the universe. This Original Motion/Long Tide of the toroidal wave-field spiral forms a dynamic equilibrium and stable ordering field that underlies and interconnects the human system with the universe, and is continually being renewed in every moment.

Our primary respiration is a stable tide within and around our body-mind, and these tidal motions arise from *AIAUch*: the original toroidal motion.[190] This Breath of Life is experienced when one is no longer identified with the analysing, separating, free-will, intellectual trinity of the mind. It appears in every sentient being as a Presence of living light and dynamic stillness (the quantum field of *AIUn*) from which all form arises.

This dynamic stillness of *AIUn* generates the Breath of Life torus field of *AIAUch*. Between the two occurs a *rhythmic balanced interchange*. In *AIAUch*, one experiences connection with all sentient beings, all of whom are held in toroidal wave-fields that enfold and maintain all life.

This is the "love" mentioned in *AIAUch*, which is experienced when one is totally in the present moment. This power of now is when one can receive and BE the love from where all creation arises.

When we experience the Long Tide arising from this original toroidal pattern of *AIAUch*, our minds and pranic forces align in *KHPHCHTHTHCATATAV*. This is the original blueprint of our body-mind, to flow with this toroidal pattern of the One in harmony.[191]

This original toroidal pattern flowing through the pranic mind is shared in Atharva Veda 19.51:

I Am undisturbed in my being, undisturbed in my vision, undisturbed in my hearing, undisturbed in my inhaling, undisturbed in my exhaling, undisturbed in my circulation and in all aspects of my Being.

This is where one Realizes that *"you are not a body and a mind moving through a world. The body, mind and world flow through you."*[192]

Prana Sounds

You can arrange the eight sounds of the Five Pranas and Trinity of Mind to compose any number of pranic-mind mantras, depending on what you wish to create, dissolve, let go of, empower, manifest, breathe life into, ignite, elevate, ascend, expand, digest, process, circulate and embody. Once you identify what you need more of, what you need less of, and what to balance within you, then you can compose these mantras.

For example, if your ahamkar is strong, balance it with Buddhi. If you wish to increase life-force, sound the Prana sound. If you feel you need to release energies, people or toxins, sound Apana. If you need more focus, structure, discipline, sound manas led mantras.

Be creative. The more you investigate and know yourself, the more bespoke the mantras can be for your unique situation.

Do each sound in each centre 36 times for a boost of energy:

KH in the heart
PH in the anus
CH in the crown chakra
TH in the throat chakra
TH in the navel chakra
CA in the left hemisphere of the brain and/or back of brain
TA in the ajna pineal chakra
TAU in the parietal lobes/right hemisphere of the brain and/or back of brain.

KA PAy: Matter and Light

परकृतिं पुरुषं चैव सर्वेषामेव सम्मतम् ।
संभूतमिति विज्ञेयं कपाभ्यामिति निश्चितम् ॥२४॥

*prakr̥tim puruṣam chaiva sarvēṣāmēva sammatam
sambhūtamiti vijnēyam kapābhyāmiti niśchitam*

24

The matter-energy code *prakritim*
the formless living light *purusam*
are One in everything;
this is understood.

Source of bliss-awareness *sambhū*,
the Intelligence Measuring out *kapā*,
forms the boundaries for everything;
This is for sure.

The uncreated universe is light undivided. The created universe is light divided into mated pairs.
– Dr Walter Russell

KA prakritim and *PAy purusam* are the fourth mated pairs of creating, the previous pairs being *AH-I* Shiva-Sakti, *EOn* Maya-Ishwara, *AI AU* Brahma-Saraswati. These four polarities and informational pattern are now fully activated and established across the whole human system, enabling an individual's embodiment to occur.[193]

PA Purusham is the living light of awareness and presence in all sentient beings. *Purusham* is living light because it is not reflected from anything else. The mind and the moon are reflections of their source, just as Maya is a reflection of its source. Purusham is living light that is not reflected off anything; it is singular and formless.

This formless presence of living light animates all living beings. As it is omnipresent, it is the same presence in all living beings, our quantum resonance. The living light in me is the same as the living light in you. This is Namaste.

Purusham is Presence beyond all qualities, attributes and experiences, beyond cause and effect, beyond time and space, beyond matter, experience and the phenomenal universe. It is not an object of experience, or an experiencer. All creation is birthed from Purusha, as the *Rig Veda* X, 90, 2, 12, 14 shares:

From his mind, the Moon was born, from his eye came the Sun. From his mouth arose the powers of fire and lightning. From the wind his breath, from his navel came the atmosphere, from his head heaven, from his feet the earth and from his ears the directions of space. Thus, all the worlds were formed... this Puruṣa is all that has been and all that is to be.[194]

The universe is this Cosmic Person Purusha; the human being is the universe and the universe is within the human being. All living things appear within this awareness. This living light shines forth in everyone, from the Sun to your best friend and your worst enemy. Purusham is *your original face*. Try and see this living light in each person and creature right now as you look around you. It is always and already here.

> *He who sees the Self in all beings and all beings in the Self has no fear. Where can there be any delusion or sorrow in whom all beings have become himself, for the knowing one who sees only unity.*[195]

In a subtle play on words, the second aspect of *purusam* is the embodied individual person: you as a *purusa*. "Each person as the individual Purusha is a manifestation of the cosmic person, the supreme Purusha, the universe as one Self. The entire universe is one Being and Person, a single reality.

We are all different manifestations and functions of that infinite and eternal Self. The goal of Yoga is the realization of that Purusha or Self as our own true nature. All divisions in humanity and in the universe dissolve in that supreme Purusha."[196]

When everything appears divine to you then you are being purusham, God Seeing Itself in everything. When you do not see the divine in everyone, then you are in limitation-purusa. In non-dual Saivism *purusa* is Shiva, yet this individual has forgotten that it is Shiva.

As Jabali Upanishad 1.2 shares: *pashupatirahamkārāviṣṭaḥ saṃsārī jīvaḥ sa ēva pashuḥ.* "Jiva is nothing but Shiva the Lord of all beings – Pasupati himself who is acting the role of egoism."

As the Maitreyi Upanishad continues, *"dēhō dēvālayah prōktaḥ sa jīvaḥ kēvalaḥ śivaḥ tyajēdaj nānanirmālyan sōhambhāvēna pūjayēt."* The body is said to be the temple; the individual self jiva purusa is Shiva alone. One should discard the faded flowers of spiritual ignorance and worship God with the conviction, "He and I are one."

Prakriti

Awareness through its consciousness creates the universe; the universe does not create consciousness. Without this awareness "matter dwells in an undetermined state of probability".[197]

KA is Prakriti: purusam's nature in action, moving, manifesting and expressing purusam. Prakriti is primordial matter-energy, undifferentiated potential in essence, assuming the flux of ever changing ever moving forms through *vikriti*.

For example, *buddhi* or intellect arises from Prakriti, yet this mind changes constantly as it is shaped by *vikriti*. All the constituents that compose the objective internal and external world are created through vikriti from Prakriti.[198]

This process shapes and forms matter of every kind, internal and external, through vikriti, its recursive fractal code. These are not just the four forms of matter in gases, liquids, solids and plasma: matter is also thought, reflected light, sound and space. Any and all movements unfold matter.

The primordial matter-energy of Prakriti is the substance of all forms. Prakriti is purusam's formless nature in action, raw matter-energy that enables relationships and interactions, shaping and forming all things through its process code of vikriti. It becomes forms as this code unfolds, repeating itself, multiplying, and forming other branches of code to create patterns in virtually infinite variations of complexity.

However, this web of code and matter-energy cannot by itself take on form and shape. Before anything manifests in creation, its blueprint is determined and its boundaries, structures, form and shape are set. "The sacred reveals absolute reality and at the same time makes orientation possible; hence it founds the world in the sense that it fixes the limits and establishes the order of the world."[199]

Purusam is this intelligent organizing presence of living light in every living being. Purusha and Prakriti are twinned, as *together they form the light blueprint and matter form of the individual.* Without Prakriti, Purusam has no way to express, no way to manifest, no way to create, and without Purusam Prakriti has no purpose, guidance or direction for its primordial matter-energy code.

Purusam awareness is the immortal, independent awareness behind your body, genetics, emotions and mind, all of which find their expression through *Prakriti.* Purusam is the architecture behind all material bodies and the material universe, which Prakriti then clothes and forms in order to express Purusam's nature.

The universe is his body, and Purusam expresses itself through Prakriti. *"Taking Recourse to my own eternal nature, I create again and again,"* as Krishna states in Bhagavad Gita 9.8. In Purusam, nothing changes, withers, dies or mutates. It cannot increase or decrease: it is always the same. There is no judgment, right or wrong, nothing to measure a self against another self. There is no comparison, no contrast, nothing to get. It goes nowhere. It just IS.

Prakriti states that we are matter and manifests an experiencer in form, bringing forth the power of relationship and interaction with others as a way to embody and Realize the One Purusam in everyone, whilst Purusam rests in formless eternal Being and Presence.

Prakriti comes to define who you think you are when you do not have a living connection to Purusam. You come to believe that you are the body-mind, and the world of matter is all there is. The matter and body that Prakriti lends us are temporary clothes, vehicles we use as individual Purusas to fulfil our experiences and karmas, giving us the means to continue our need for experience *until* we are satiated with worldly experience and come to feel there is something more.

In this understanding, we open up to what is beyond matter, the body-mind, the world, and indeed experience itself: the underlying reality of what never changes, and what has never been formed and created.

Prakriti can lead us to this infinite peace, freedom and presence of Purusham if we look at our experiences from the perspective of Purusham. This is Realized when one is not identified to any form or change that Prakriti lends us. Prakritim and Purusham work together as this is their purpose: to clothe the formless with the matter-energy of form.

The Code

Prakriti's change making agent vikriti is its fractal, recursive algorithmic code, which brings forth our ability to subjectively experience life and relationship to other bodies as an individual who has form, who changes, grows, withers and dies. This code creates moving changes, which is the nature of the world and all experiences, manifesting the virtual reality hologram of the universe.

Fractal Science as demonstrated through the Koch snowflake and Stephen Hawking's wisdom has discovered that there is a limit to the space, time duration and area of this virtual reality simulation, *but* there is no limit to the complexity of the simulation and our experiences within it. There is potentially no end to having, and being identified with, experiences and forms generated by Prakriti.

In other words, the movements and changes of our life experiences can become more and more complex, as we see in the increasingly complicated world of today, and this can go on virtually forever. Our experience within the simulation can be virtually infinite, but the space and time duration of the simulation is finite.

Change and growth in the simulation experience levels out at around 1.618 – the number of the Golden Mean, a virtually infinite spiral coming from infinite time and echoing into infinite time. The fractal information of the simulation pattern (vikriti as the agent for the raw energy matter of prakriti) repeats itself on all scales. To change these patterns in your subjective life entails you reaching outside this simulation experience of prakriti-vikriti to that which is *not* identified with the simulation: Purusham.

This part of you Purusham is not identified with or bound by the laws of the virtual reality simulation of the universe prakriti-vikriti. This virtual simulation is a fractal code of information structured via a complex algorithmic code. We exist *outside* this reality simulation code as Purusham, as well as *within* this reality simulation code as the body-mind energy of matter prakriti.

We can commune with the part of us – Purusham – that is out of the virtual reality simulation through meditation, awareness, prayer and evocation of this pure awareness through an awakened teacher of non-duality. Purusham can help anchor and guide us through non-duality whilst we are in the simulation so we can learn our lessons about the nature of the simulation of duality, and not get lost in it.

Moving our focus and awareness into living That which is beyond the simulation is the goal of all sacred traditions. That which is beyond identifying with the simulation as the only reality is our original environment, our base line fundamental reality: freedom or *moksha*.

So: Purusham projects an image or avatar Purusa, which is clothed into form through Prakriti, giving it substance, informational structure and code to create a fractal recurrence on multiple levels of complexity. "Man is made in God's Image" and "Ye are like Gods" are two Biblical expressions that spring to mind.

Each Purusa's or soul's matrix is made of vibration and reflected light: the primordial matter-energy of Prakriti, working in a dual relationship between 0 Purusham, its light blueprint and 1, the matter energy of Prakriti. This matrix is the pod or home of each purusa soul, from where it incarnates and reincarnates back into. As it reincarnates, it brings this information back to its source-pod, OR if it is not reincarnated the soul pod matrix that Prakriti forms, dissolves.

The purpose of this is for the One to experience itself in an almost infinite number of variations and complexities, until the time duration of the simulation established at the creation of a universe is completed.

Then, this universe dissolves into a "Big Crunch" and a new universe appears in another "big bang" with a new set of parameters, laws and sentient creations to experience a new creation, which then resonates all this information to the ever connected One that is experiencing this, yet is not identified with this experiencing at the same time.[200] Consciousness communicates to itself all the time, informing itself through its creations, experiencing the many facets of Itself through its sentient creations in an ever-connected feedback loop.

Similarly, we inform ourselves through our creations. We learn by creating something, whatever it may be. Teachers learn from teaching, artists learn by manifesting their art, writers learn by writing, and healers heal themselves by helping heal others, hence the term "Healer heal thyself."

Having a child is a common experience where we get to see ourselves in miniature form and are informed about different parts of ourself which we previously would not have experienced. We become informed of our own self and our own creations through our children in an ever-connected genetic feedback loop.

This is designed to lead to a deeper love within us, as a reflection of how the One created us as its Children, showing us that the unconditional love we share with our children is how the One Purusham loves each of us as individual purusas, and indeed is the love that we are.

When the personal experiencer of purusa is no longer identified with, your existence is guided by the degree of Presence you have in your daily life. With the guidance of Purusham you can live the quantum Code of the One and remake the simulation in your own image, thereby mastering the simulation. Established in Purusham, one can enjoy Prakriti for all its worth.

And this is the reason why we are in the simulation – to learn from it and to master it in order to realize oneself as Purusham, not just to repeat Prakriti's virtually infinite complex patterns over and over in many lives.

Quantum Computing

Scientists are developing the next generation of quantum computers by harnessing quantum laws which work with both light and matter: Purusha and Prakriti. These quantum computers have light speed processing power that can power Artificial Intelligence and Virtual Reality engines in order to create worlds (like Maya designs).

Quantum computers replicate Purusha and Prakriti, the two fundamentals that design and form the universe and human being. In quantum computing, information is stored and processed in 0s, 1s, and all values in between – rather than just the clunky and dualistic 1s and 0s of classical computing.

The information energy of matter 1 (Prakriti) travels at the bridging speed of light, where information is stored and processed in 0s, 1s, and all values in between (Gha Dha Dhash) with Purusha being beyond the speed of light 0 as quantum light or living light.

Scientists in 2022 have successfully produced the largest hybrid *particles of light and matter* ever created. These *quasiparticles* or *Rydberg polaritons* were made with the help of a crystal which was thinned to less than the width of a human hair and sandwiched between two mirrors (two reflected polarities within a unified field) to trap light. This achievement could create a quantum simulator that runs off these Rydberg polaritons, using quantum bits – *qubits*.

"Making a quantum simulator with light is the holy grail of science," says physicist Hamid Ohadi, from the University of St Andrews, UK.[201] "We have taken a huge leap towards this by creating Rydberg polaritons," which *switch continually from light to matter and back again.*

Light and matter are two sides of the same coin; light particles move quickly, but do not interact with each other. Matter is slower, but is able to interact. *Putting these two together defines, forms and creates the universe and the human being,* and one way we can replicate this ourselves is through quantum computing.

This dance of matter and light includes all possible values and relationships between 0 and 1, and is a defining characteristic of the quantum universe. Understanding this scientifically *and* experiencing this through expanded states of consciousness will help us define and guide these quantum states, so we can become quantum creators ourselves.

However, it is important to remember that the quantum universe remains undefined *until it is observed* (EOn). This is also why the Maheshwara Sutra is crucial, as with its wisdom we know we can observe-project-create as Purusha from the undifferentiated raw energy information of Prakriti.

Harnessing this ability can lead into limitless possibilities of being able to create forms of any and every kind. The potential is for humans to become like gods if informed by purusha, or to become like demons if solely informed by prakriti and maya: we can design heavens or hells, we can design illusions or reveal the veils to illusions.

Embodiment

God has to be half mortal and half immortal.
– Satapatha Brahmana

When one overidentifies with Prakriti in alliance with Maya, the architect of measures and reflections, one can become mired in and overly attached to your body-mind and the world of matter, experiences, forms and appearances: *samsara*. One can always be seeking in the world for fulfilment and happiness, and the search will be virtually endless in the virtual reality simulation.

However, Prakriti points one towards what one needs to experience and learn in life IF one understands and guides Prakriti's promptings through Purusham. For example, if one desires an intimate relationship with another believing this will fulfil you, you are seeking outside yourself for fulfilment, and this will never lead to Purusham.

However, if you are happy to be alone and desire a relationship in order to know yourself better, to see parts of you that you cannot see alone, to evolve, this can lead to Purusham. The need for relationship is a human reflection of Shiva-Shakti and a fundamental human driving force. Prakriti is all about relationship: Purusha has no relationship to anything, for only the infinite can know the infinite.

The need for relationship is only fulfilled within a sense of one's own sovereign Self or Purusha in the whole process. One needs to be consciously aware of Purusham and its supremely detached pure awareness-love in order to cognise the workings of Prakriti, and to then work with it mindfully in your life.

The promptings of Prakriti to experience ever more complex patterns within the simulation can be seen clearly for what they are by the awareness of Purusham. This helps one to not blindly follow these needs and attachments and become lost in the material world and its ever-changing experiences. When one can see Prakriti's impulses and workings clearly for what they are, one can complete one's karmas mindfully without much stress and attachment to them, *or* one can leave these impulses totally.

The fulfilment or dying off of these impulses within yourself allows one to rest in the Presence of Purusham. Once your needs for certain experiences have been seen for what they are, they can manifest in your worldly experience, or not. The best outcome for you will happen in surrender to Purusham, because everything arises from Purusham.

Purusham states that you are not a body, a mind or an experience, but rather the awareness in which these appear, and with which they are Known. Awareness does not share the limits of the body-mind.[202]

Similarly, exercise can help you embody and live Awareness if done beyond being identified solely with the body-mind, and with the awareness that the body and mind appear in Awareness. For example, good yoga teachers will include Awareness in their yoga practice. Similarly, when we are one with the experience of running and at the same time free of the experience of running, this too helps embodiment.

Dancing can be a way to break down armour and enter awareness in the flow. Appreciating any specific form can lead one into formless awareness if done mindfully. Lovemaking can be a journey into formless awareness in the present moment through our bodies.

Embodiment involves becoming sensitive and aware of the body and its communications. The body of Prakriti is a barometer. Embodiment is being present to your body, your feelings *and* the living light of awareness. This asks us to be light and matter – embodied humans.

The awareness of Purusham, the indwelling spirit, is the ever-present Presence in which the body appears. We are not bringing something down from heaven to be here in us, and there is nothing to call forth from some external agency to bring into us. Purusham is already and always here.

One part of Realizing Purusham with our embodied self Purusa entails that we have to clear shocks, emotions, traumas and their grooves of conditioning from within us. All of these are not part of the quantum Code *and are unnatural*, so we have to dissolve, release and embrace everything that is not This from our *cellular memory*.

This is a key way to embody Purusham. The substance of these conditionings is made of Prakriti, and without this release of repressed emotions, traumas, genetic conditioning and other subtle substances, Realizing Prakriti-Purusham and embodying this is almost impossible.[203]

True embodiment means we have healed our wounds of incarnation and fulfilled our unique menu of earthly lessons and experiences. These experiences manifest in the embryo at *EOn*, and include karma from our past lives, the imprints of our natal time dwelling in our mother's womb receiving our parents' imprints (which are our own karmic reflections), the influence upon us of our environmental imprints (school, friends, relationships, teachings, jobs, teachers), heartbreaks and emotional traumas. All these experiences influence our choices and directions in life.

Beings who have done this totally, like Sri Aurobindo, the Mother, Christ and others, embodied Purusham so totally that their physical bodies mutated from the standard bodies that most humans have today, with parts of their "junk" DNA being activated by this living light.

This is taken to its height in the phenomenon of the Rainbow Body, whereby upon a person's death the body dissolves into light, leaving just a few hairs, teeth and part of the skull. Various masters, Siddhas and Buddhas have attained this Rainbow Body, as did the Apostle of Christ St John.

One night, upon his request, he was placed into a hole on the island of Patmos, Greece. His followers were astonished to find that his body had disappeared in the morning, with not even a single hair remaining of his body. This mastery of Prakriti occurs when the living light of Purusham is fully cognised and the individual Purusa lets go of its identity.

Our body is a precious gift as it enables us to live, learn and bring living light into play. "Form is emptiness, emptiness is form," as the *Prajnaparamita Heart Sutra* shares. When we experientially Realize the formless living light within our cells, then Purusham and Prakriti live and embody their Union in us.

Self-Inquiry

Are you experiencing the peace and freedom of your cosmic being at odds with your material responsibilities? What aspects of embodiment do you need to develop? Why aren't you doing this? Are you a victim of circumstances and life, or a director?

To fulfil our soul purpose and our karmas, we must overcome obstacles within us. A healing of a wound, a change in perception, a different way of communicating, can all work wonders in helping you achieve this.

For some people, this may mean diving more into their body. For others, it may mean less bodywork. For others, it may mean doing yoga and exercise with awareness: for others it may mean doing less exercise as an escape. Everybody is different and has different routes to embodiment.

Prakriti, Vikriti and DNA

If you think you are enlightened, go spend a week with your family.
– Ram Dass

The undifferentiated energy matter of Prakriti unfolds and shapes its many forms through its code of vikriti, giving us an opportunity to play out our karmas and learn from our life experiences. Our DNA is an expression of this code of vikriti-prakriti, which forms and shapes your individual self or purusa.

Our DNA contains our ancestral and family memories, emotions, traumas, patterns, beliefs and gifts that offer us opportunities for recognising forgotten, fragmented and wounded aspects of ourselves.

Underneath the streams of our lives lies a river of the experiences of our ancestors. We are a "web" of these experiences, and until we release them we cannot fully access Purusham. Our ancestors have both positive gifts to share with us, and negative traits. Many of their positive traits are not passed onto us fully because of our parents' and ancestors' inabilities to do so, and because of our own blocks.

We chose our parents and their bloodlines as a mirror for our own karmic tendencies in order to facilitate experiences for our own growth and freedom.

Our DNA is also a recorder of our physical, emotional and mental developmental patterns, as well as being a recorder of significant past life actions and experiences, as well as past life abilities and gifts.

Our DNA stores massive amounts of information in a very small space and is not just physical; the DNA helix extends itself through vibrating superstrings from our bodies to weave throughout the earth's grids and outwards into the whole matrix of creation.

These superstrings constantly emanate vibratory signals, messages and transmissions throughout your body-mind, electromagnetic field, and outwards into the earth's fields and beyond, *all the time*. This creates your own genetic matrix, genetic field and genetic grid around you and around the earth that informs your perceptions of reality on a subconscious level until you become conscious of it.

There are 7 layers to this genetic programming:

1. Your mother, father and current family
2. Your ancestors from two generations ago up to 14 generations ago
3. Your Lunar and Solar Ancestors
4. Your Stellar origins
5. The Progenitor Rishi who is the founder/head of your Bloodline
6. Manu/Brahma: The first human parental DNA pattern arising from the Creator Intelligence
7. Purusham

Becoming conscious of these layers enables you to release crippling genetic patterns and ancestral traumas, which greatly benefits you, your family, your children, your ancestors, and the entire gene pool of humanity.

The process begins by clearing the traumas, beliefs, patterns and limiting behaviours you have taken on from your parents. This then extends into excavating patterns, emotions, beliefs and traumas for 2-14 generations before you; for most people, it will be between 4-7 generations previous to you.

The next step extends into your solar ancestors, known in India as the *Adityas* and *Vivasvan*, and then your stellar ancestors, the star seed race you came from to earth, the star you feel most aligned to. In stepping more into your Self, you meet the Head of your Bloodline or *gotra* which in India comes from one of the Seven Rishis, the progenitors of much of the human race.

Most Indian people and many of the peoples of Asia, Eastern Europe and the Indian subcontinent have one of the seven Rishis' genetic imprint within their DNA. Your personal Rishi or "Great Father" can help you in many ways to fulfil your family/ancestral karmas and duties. It is very useful to know who this is, as he will be open to helping you in profound ways.

The sixth level of our genetic conditioning is Manu and Brahma. In India, the first human parent is not Adam or Eve but *Manu*, the original ancestor of Vedic humanity. This Manu is created in every great cycle of time or *manvantara* by Brahma in order to populate the earth.

The seventh level is *Becoming Purusham, which involves two births*. Our first birth is through our mother's womb as a purusa individual self, formed in matter by Prakriti. The second birth is when you become free from all conditioning, and is called *"dwija"* or *becoming twice born*. This happens with your Self Realization and dissolution of your conditioning and genetic imprinting, leading to a neurobiological mutation or switch being turned on inside you.

The quantum Code is realized within you. This Code is then bolstered in the collective human gene pool, acting as an accelerator for evolution. This in turn makes it easier for others to release their own genetic and ancestral imprinting and to become twice born.

You can only Be Purusham when you are not *being* anyone else, or unconsciously aping or trying to be like someone else. This includes acting out programs of behaviours, wounds and traits from your family and ancestors, culture, peers and partners *on all levels*.

You cannot truly be yourself with ancestral burdens and unresolved wounds running through your emotions, body and mind. You cannot be yourself when you are still being your parents and your parents' parents, and neither can your children be themselves fully either.

To be your natural Self Purusham is the greatest gift you can give to yourself and the world. This Sovereignty is Independence, *sva tantra*, Self-government and Self-authority over all aspects of yourself. In living this, you exist at the cutting edge of evolution, transmitting signals that give all others permission to be their natural Self too. You become a Light Unto the World.

Seven Layers Genetic Programming

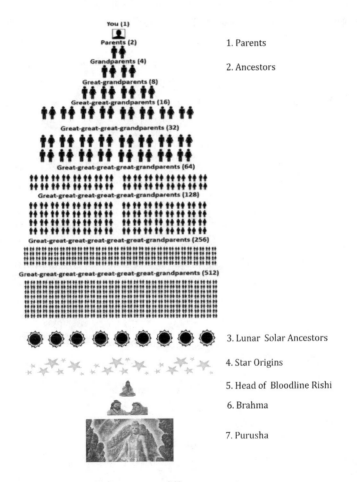

1. Parents

2. Ancestors

3. Lunar Solar Ancestors

4. Star Origins

5. Head of Bloodline Rishi

6. Brahma

7. Purusha

The Evolution of the Gene Pool

The human gene pool holds the collective genetic information and total number of *genes* available in our *species*. The gene pool is sexually powered, and is the total "breeding stock" available to our species. Our own personal gene pool comes from our ancestors and parents passing on their genetic information to us, creating an ancestral consciousness history resonating within us that emanates its information constantly.

Your parents and ancestors have deep patterns that play out through your DNA and your children. In esoteric lore, every bloodline needs at least one person to catalyse and initiate a system-wide defragmentation or healing of the lineage. Many of the more conscious people on earth, like you, have chosen this role out of unconditional love for your parents, your ancestors, your (future and present) children and of course yourself.

We are the ones on the "frontline" of the "Family Story", carrying in our DNA the endurance and suffering of the generations who have gone, and those still to come. We are now healing and closing these timelines. It is our mission to ensure "It Ends Here".[204]

Genetic healing is a visceral physical, emotional and spiritual experience. It changes the DNA, brain and body-mind, and you can feel it deeply in your body and emotions. The experience of feeling your ancestors with you, within you, and around you, can be deeply catalysing and touching. Feeling a deep connection to your lineage in loving humility, sincerity and purity, is honouring and profound.

In the release of genetic imprints, your ancestors, family and children, both present and future, also receive impetus for their own evolution and healing. This release impacts your future too as it changes the possibilities available to you, shifting your potential future possibilities from one outcome or timeline to entirely new timelines and possibilities.

Your future timeline changes as you heal, and your DNA increases its vibratory rate the more it is released from its layers of ancestral and familial burdens. (These burdens are part of your own mirror and your own learning.)

What may seem improbable or even impossible *right now*, will suddenly become do-able, and more than that, will tangibly manifest in your life and relationships. This in turn creates a whole new set of possibilities and future outcomes.

When you take away all conditioned emotions, mind, genetics, beliefs and patterns of conditioning, what is left? When you take away all experiences of change, time, space, karma, birth and death, what is left? Purusham.

Purusham is the formless living light blueprint of the Code of Life in your DNA, its source Presence. Imagine: every person, animal, plant, tree and life form are within you now. Look around you. Anything alive is literally within you at this very moment!

All in all, *KaPay* marks the penultimate coagulating and embodying of awareness and primordial matter-energy into form. This stage in the development of the foetus marks the embodying of awareness in the individual body-mind-self, and is the Knowing that you are more than the body-mind *purusa*, and are in fact *Purusham*. The body-mind and its matter is formed out of *prakritim* as its primordial energy substance.

Sounding Ka Pay

Sanskrit truly is a language of mantras. Each syllable evokes different aspects of The One. With intelligence and intuition, we can guide ourselves into the aspects of the One we need to align to.

KA is the syllable for Prakriti. This is a semi-guttural minimum breath sound made by a contraction in the throat with KA, mouth open, tongue relaxed.

You can sound this at the base of your spine. Sit with your hands forming a square in your lap. Sound it x54 aloud, then x54 silently.

PAy (pronounced like "pie") is the syllable for Purusham in this context. This is a labial (lip) sound with minimum breath, with tongue relaxed. Purse your lips together and blow out PAI.

You can sound this at your crown chakra *and* at the soul star chakra, one hand's length above your crown. Sit with your palms open on your lap, or with one palm over the other. Sound it x54 aloud, then x54 silently.

You can add specific intentions when sounding KAPAY. What do you need more of? What is your weaker area? Do you need balance between the two? What are your incomplete needs, desires and experiences you wish you had, but never got to fulfil? What aspect of Purusham and Prakritim do you need to engage more or less with?

Bring these into the sound meditation as an intention/prayer/sankalpa you make with the sound.

- You can unite *KA* and *PAY* with 108 repetitions on the crown chakra with *PAy*, and then switching to the root *KA* for 108 repetitions.

- To access more Purusham, you can add *RLRK* at the ajna chakra third eye, and sound *PAy* at the crown chakra.
- To access more specific qualities of Prakriti you can add the specific three Guna sounds you need to (see next chapter) with *KA* at the muladhara root chakra. You can also add the syllable *LAN*.

SHA SA SAr: The Three Gunas

सत्त्वंरजस्तम इति गुणानां त्रतियं पुरा ।
समाश्रतिय महादेव: शषसर क्रीडतिप्रभु: ॥२५॥

sattvarajastama iti guṇānāṁ tritayaṁ purā
samāśritya mahādēvaḥ śaṣasarkrīdati prabhuḥ
25

The three qualities of light, movement and stability
evolve prior to the creation of the worlds,
illumining creation in the ever-rolling wave of play *śaṣasar*.

The God above all gods *Mahadevah* constantly presents Itself in
all embodied beings.

Sounding *śaṣasar* in 1 beat,
these three gunas cycle, move, multiply and exist
in every thing.

In *Mahadevah* is *Prabhu*,
Master of creating-in-action
yet ever the Non Doer.

शकारादरजसो भूतिः षकारात्तमसो भवः।
सकारात्सत्त्वसंभूतिरिति त्रिगुणसम्भवः ॥२६॥

śakārādrajasō bhūtiḥ ṣakārāttamasō bhavaḥ
sakārātsattvasaṁbhūtiriti triguṇasaṁbhavaḥ
26

Rajas is the process of action, and originates from *Sha*.
In completing actions comes *tamas*, forms and objects, which
originates from *Sar*.
In light sattva, *Sa*, you come closer to *Sambhu*.
The origin of sattva is from the letter *Sa*.
Sambhu generates the three moving, stabilizing, light forces
that interplay within everything.

The totality of manifest creation, extending from the unmanifest to the full extent of every created thing, is held by the grip of these three fundamental forces.

– Enlightenment! The Yoga Sutras of Patanjali by Maharishi Sadashiva Ishaya

From the beginning of creation *A I Un* comes the trinity of Light *A*, Action *I* and Existence *Un*. This is reflected now in the trinity of SHASASAr: *SHA* Action, *SA* Light and *Sar* Form. In between the mirror of these two trinities lies all manifest creation.

The trinity of SHASASAr are called the three *gunas*, and they move throughout all of creation. Guna means a quality, a constituent, a thread of a rope, a string and a division measuring out part of a whole. Guna implies vibrating, repeating, reiterating and multiplying itself. The three gunas interact throughout all creation as light manifesting moving waves of vibration into forms. *Nothing could exist or be experienced without these three.*

The three gunas show us the three ways a wave operates. Rajas *Sa* is the masculine, moving action of the wave, *Sha* is the neutral light, and tamas *Sar* is the feminine spiral bringing the wave into form, collapsing the wave function. *This Law of Three* is seen throughout all life's processes, as for anything to manifest anywhere it takes a combination of three different forces to do so.

Each guna is seen as a thread or string with each string combining with the others to form a "rope" or toroidal spiral. These strings consist of the centrifugal, ascending positive rajas wave of action and its polarity, the centripetal descending negative spiral of tamas organizing this action into forms. The central sattva is the neutral light holding, illuminating and informing them both, keeping them aligned and in balance when consciously adhered to by the individual.

The gunas are cycles of change within a continuum, as spirals expand, contract and can move in all directions simultaneously, from its extremities to the centre and the periphery, in one fluid interchangeable flow.[205]

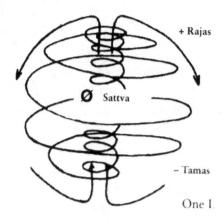

In a perceptual sense, "you" as the seer or subjective I is sattva, that which is seen or perceived as objects are tamas, and the energy of perceiving between subject and object is rajas. Any movement the self makes involves these three.[206]

The three gunas govern our actions, reactions, responses and our lack of actions in life. All three intertwine to manifest a sense of self and other, guiding these perceptions. They intricately interconnect as the basic energies of Nature *Prakriti*, combining and re-forming in virtually innumerable permutations in every manifest aspect of creation.

The guna you are most drawn to is the way in which you predominantly act, feel and think. One guna will dominate your moods, actions, emotions and behaviours. Reflect on this as you read on.

Sha: Rajas: Action

Rajas evolves from the sound Sha, and is the mirror of I, the second sound of desire, movement and action in the quantum field of creation $A\ I\ Un$. In this sense, the feminine energy action I is the origin of the masculine action *Sha*.

Rajas is a positive wave of vitalizing, creative kinetic energy. It activates and catalyses action as a centrifugal spiral of masculine energy flowing from the manipura solar plexus chakra. Rajas is the force of action that rules and generates motion.

In rajas we are action men and women. We have vital force, we move, we are dynamic, passionate and energetic. Rajas moves outwards and pushes to get things done. This busyness can be energizing or distracting, creative or destructive, for where rajas goes and what it does depend on whether you incline towards sattva, light, or tamas, darkness.

Rajas is a directional creative force, exciting what it comes into contact with. It is fiery and is usually focused on the external world, from where the ego *ahamkar* gains its sense of achievement and self-worth. When rajas is focused and directed inwards, it can be harnessed as a powerful force to deepen your meditation and sense of true self-worth.

Rajas is the masculine *pingala* channel that weaves it way around the right side of our spine, and it governs the mind of manas, the active doing mind. When unbalanced, this becomes the monkey mind, constantly overthinking and babbling away.

When we have too much unbalanced rajas we become unstable, constantly chopping and changing our focus, plans, schedules, ideas, actions and thoughts. This leads to a sense of unreliability and uncertainty.

Constantly shifting and ever changing rajasic people can be blown around like a leaf in the wind at any moment, as they have a lack of grounded trust in their bodies (tamas) and not enough harmonious light sattva to guide them into peace and contentment.

Rajas generates our thirst for stimuli *and* for experience. Unbalanced rajas is never at peace or at rest; it is restless and always searching for something to do. Rajas sustains life actions, yet it can also entangle us into the matrix of *doing-doing-doing*. It can limit us when we become attached to action, experience, stimulus, doing and being stimulated.

Rajas is motion. Motion connects one event in space-time to another, creating a narrative for the self to navigate, coordinate and function effectively. Motion connects subject to object, the physical to the mental world. Rajas is the perception-energy-motion *between you and objects* (remembering you are sattva or the subject and all objects are tamasic), the energy between you and another person/object, *the energy of relationship in a self-centred context.*

Overly rajasic people can come across as arrogant and bossy. Their vices of lust and greed fan, and consume, their inner fires. They can be impatient, fast to do everything, quick to anger and judge others. They can be defensive, with hard walls giving off a false air of invulnerability.[207]

Rajasic people can also be overly optimistic and therefore create illusions. They like to be independent and are compulsive, addictive types. Passionate and sexually active, rajasic people are fast in talking, doing and moving. They can be catalysts, motivators and inspirational figures like life coaches/mentors, as they can move energy and obstacles fast.

Tamasic or more inert people like being around rajasic action people as they give them hope, inspiration and energy. Rajasic foods include meat, sugar, coffee, fish, garlic, onions and spices to stimulate you, to give you a boost of energy.

Reflect on what 3 things you need to do, and stop doing, to balance your rajas.

The deeper understanding of this verse in the Sutra is about transcending rajas, and is given by the female sage *Sri Anadamayi Ma*:

> There is a stage where working is delightful and gives intense happiness. Here one is quite unconcerned with what may or what may not result from one's action; the work is done entirely for its own sake, for the love of it. Work for the sake of work, work itself being one's only God. If one goes on performing action of this kind, there comes a day when one is liberated from action. It is a type of work not actuated by desire or craving, one just cannot help doing it. When God manifests himself in the form of some work, which therefore exercises intense attraction on a person, then by engaging in this work again and again, one is finally liberated from all action.

In other words, "If you become so completely concentrated in any one direction that you cannot help acting along that line, wrong action becomes impossible. In consequence, action loses its hold on you and is bound to come to an end. One cannot act wrongly. The human body has entered a current of purity and as a result *satkarma*, action in harmony with the divine will, occurs."

This is the refining of rajas as it becomes an instrument of divine will and *sat karma*. Yet before this wonderful state occurs, our unbalanced rajasic actions create more karmas, which in turn arise from our *bhavas*, our inner state, sentiments, moods and emotions.

If your *bhavas* are caused by worldly events, people and the mundane, they are full of the happiness and distress, the ups and downs of life, which arise from the three gunas. But, if your bhavas manifest in relation to God, *not from the external world of experiences,* they become transcendental and full of bliss *ānanda,* not subject to the three gunas. In its height, bhava is soulful ecstasy, usually generated from deep, one pointed devotional feelings towards God.

In spiritual practices, we attain different states of bhava or happiness, bliss, peace and well-being, which whilst useful on the path, does not lead to enlightenment *as it is still just a state.* If there is still a desire to be in this blissful state, which inevitably comes and goes, taking you up and down, ascending and descending, that too is a play of the gunas. Only when your bhavas, your inner moods and dispositions, which lead to actions and karmas, are cognised and let go of, can one go beyond them.

This is the journey. Bhava is the fuel for actions to become karmas. Bhava is the seed of latent karma, and karma is bhava actualized.

Sa: Sattva: Light

Sattva evolves from the sound SA. Its mirror sound from the first trinity of creation *AIUn* is *"A"*, the first light of creation. *Tva* means "the true you", and when added to *"sat"*, which means truth, Being and Presence, sattva means *"the true you of being, presence and light"*.

Sattva is your conscious, mindful self in harmony, peace and goodness, compassionate, light and happy. Sattvic people are friendly, joyful, open, equanimous, good tempered and even minded. They are graceful and attractive to the soul.

In the words of Patanjali in the *Caraka Samhita*, the classic textbook of Ayurveda, "persons having sattvic essence are endowed with memory, devotion, are grateful, learned, pure, courageous, skilful, resolute, free from anxiety, having well-directed, serious intellect and activities, engaged in virtuous acts."[208]

Sattvic dominant people are content with what they have, are honest, and live in a state of flow. Balance and moderation are their bywords, and they are forgiving, respectful people. Sattvic people are empathic and caring: they feel others and are receptive. Compassionate and nurturing, they are supporters of people and serve others. Light is their countenance, and they have a peaceful aura attractive to others.

There is little suffering in Sattva, for Sattva is desire-intention refined into virtuous, life-affirming actions. Patanjali states that when one has fully cognised Sattva, and has thereby transcended sattva, one gains authority over all that exists, and gains all knowledge.

As beautiful as this sounds, there are traps in the sattvic state that limit our evolution. We may misinterpret Sattva and become attached to happiness, to being pure, righteous, and even to knowledge itself. "Spiritual bypassing" is a good term to describe what happens when we ignore the grief, trauma, pain and shadow within us by constantly reaffirming our good side and avoiding our pain.

A modern day sattvic escapist theme is where everything is labelled as "love and light" and positive "affirmations" are used to escape the miseries of one's existence. Sattvic people can idealize and use universal truths about love to avoid feeling themselves, and thereby get stuck in their evolution.

Mythologically, the demigods of Sattva are in a state of supersensory enjoyment in the heavens. Being satisfied with their heavenly abode, pleasures and enjoyments, they do not feel the urge to examine or improve themselves. We also do this in many ways, some subtle and some not so subtle, crafting cunning physical, emotional, mental and spiritual strategies to stay "happy" in our bubble, avoiding seeing or feeling anything unpleasant.

The beginning of the Buddha's life, where he lived in sattvic splendour shielded from the sufferings of the world, poverty, old age and death, illustrates this well. Once we let go of the attachment to the Sattva quality, we become a Bodhi-Sattva, or Awakened Sattva: we go beyond sattva but still can utilize sattva (without identification or attachment). Sattva is not enlightenment, as awareness (Shiva) is free from sattva too, just as awareness is free from all the gunas.

Sattva is connected to the central channel of white light *Sushumna* in the spine. Sattva is this illuminating light that is neutral, beyond masculine and feminine polarities. It keeps the polarities of tamas and rajas, the ida and pingala channels snaking around the left- and right-hand sides of the spine, in harmony with this light.

Sattva is a gateway into awareness, as a sattvic person works more with the inner self and does not react much to the ups and downs of the external world. They internalise reality and master themselves more. Humble and self-responsible, they are self-actualizing: they live and manifest who they are. They know and value themselves *and* their gifts, talents and capacities.

Sattvic foods are high in vibration and life-force, such as fruits and certain super-foods, and lead to a lightness in body-mind and spirit, as these foods support meditation. The more Sattvic you are, the more you will be drawn to these foods.

Ways to increase Sattva are through meditation, chanting, selfless actions, compassion, refined breathing techniques, eating the right foods and drinking alkalized water, increasing your intake of higher frequency energies, pranas, people and vibrations.

Reflect on what 3 things you need to do, and stop doing, to become more sattvic.

Sar: Tamas: Stability

The sound that tamas evolves from is *Sar*, the mirror sound of *Un*, the third sound of the quantum field and first trinity of creation *A I Un*. *Un* forms the all-pervading sense of existence itself, the quantum harbinger of form.[209] Tamas *Sar* brings this *Un* existence into solid form.

Tamas is a centripetal inward pulling spiral, the feminine polarity of rajas. It is the spiral of entropy, a contracting, decaying wave that collapses the rajasic wave of action into form. Tamas is known as the perisher, the darkness, the decaying force, and death. It is as necessary to creation as rajas is.

Tamas forms structures and boundaries that help collapse the wave function into form, so that forms manifest. All alchemy happens when a container is created, when limits and boundaries are established, which allows forces to coagulate and manifest.

Tamas is the force of gravity, that which holds the planets together whirling in synchrony around each other, and tamas is the glue that binds atoms together. Tamas is stability and density, the force that makes everything solid and tangible. It gives rise to the physical body and death, and is part of the power of time, or *Maya*. No forms or bodies can exist without time.

Tamas is a force of contraction that can be useful and harmful. It collapses and contracts the wave of rajas into forms, enabling us to have a world of objects and people. All objects are tamasic. Tamas helps form the sense of *ahamkar*, the separate self, and we need this boundary in order to exist and navigate in the world. However, with too much tamas we get fixated on objects, believing that the 3D world of objects is the only reality.

Unbalanced tamas manifests as resistance, stuckness, fear, laziness and decay, as well as the fear of death. Through sloth, selfishness, narcissism, overeating and oversleeping, tamasic people do not do much of anything, becoming a literal "couch potato" with little motivation, action, inspiration or desire.

Tamas rules negativity and the reactive, unconscious, dull, un-inquiring, uninterested, rigid, fundamentalist static mind. Too much tamas leads to a denying of life and love, as hatred and anger are tamasic qualities. Having rajasic sex, sex without love, without care, leads to tamas, as our vital force gets drained. Too much tamas makes you hesitant, doubtful and ponderous in your mind, movements and body.

Mythologically, tamas is the quality of the demons or *asuras*, whereas sattva light is the quality of the *devas* or gods. The demons of tamas are said to be so steeped in ignorance that *they are not even aware of their own existence*. When tamas is distorted without the guidance of sattva, it can lead to cynicism, narrowmindedness, fear, fundamentalism and rigidity, distorting our sense of groundedness, stability and presence with the body-mind.

Overly tamasic people are contracted in many ways: emotionally they may be holding onto repressed and heavy emotions such as grief, guilt and shame. They may be overly critical of self and others, have low self-esteem and feel unworthy and self-doubting.

Tamasic people often live in the victim mentality, haunted by their own inner judge, critic and persecutor. They are often discontent and pessimistic, resigned to their "fate" or karmic malaise, unable to rouse themselves into action. It takes them a long time to achieve anything.

When one is overly tamasic, depression, judgment, resentment and hatred of self and others can arise. One can become codependent, needy and possessive, as despair is a tamasic companion. Tamasic people have poor *or* rigid boundaries, and may crave attention and vivacious people as an energy source or stay away from them completely. Tamasic people may be overly fragile and vulnerable all the time, frozen in their expression, unable to get positive action or motivation occurring.[210]

When a person has too much tamas over time, they cease to care about others or themselves as they become overly absorbed in their own subconscious self. They become heedless, not listening to others or even themselves. They live in the unconscious, guided by the unconscious, living the mundane and basic obligations of life, stuck in humdrum routines like a zombie.

Tamasic foods are low in vibrational frequency and prana, such as junk foods, processed foods, stale foods and meat. Intoxicants of all kinds are tamasic, and tamasic people do not do much exercise and are generally unhealthy.

In the body, tamas is connected to water yet is ruled by the earth element. It manifests as the feminine *ida* channel weaving around the left-hand side of the spine, and is connected with our exhalation as a cooling energy. Tamas balances too much rajas, and can relax rajas and bring rest, yet ultimately this is best done with sattva involved as well.

Tamas can help us ground ourselves into the here and now, to stabilize ourselves, to become like a solid tree or pillar of strength and reliability *if* we use it in harmony with the other gunas. Tamas helps us to embody ourselves, to be steady and unwavering in the midst of all that happens in life.

The full embodiment of tamas is the Cosmic Serpent Sheshnag – the stability that supports creation into being. Even today, before commencing construction on a building, many builders in India ceremoniously lay a foundation stone that represents Sheshnag.

Reflect on what 3 things you need to do, and stop doing, to balance your tamas.

Self-Inquiry

What guna are you predominantly operating on?

Is something missing in each guna's expression in you?

What do you need more or less of in each guna?

Do you need more of the qualities of one guna than the other?

You can change some of your vibrational patterns by adding more of, or subtracting some of, the different gunic qualities, through your conscious choices and actions. If you need to meditate more, change your diet, do more action, rest more, be more grounded, be kinder to others, have more focus, or be more relaxed: all these actions will affect the gunic balance within you.

Within you, in honest self-inquiry, you will come to know which actions to take and which actions to stop taking in order to balance the gunas, by seeing which guna in you needs which ingredient.

Even though each guna has its unique qualities, they are all intricately intertwined to form one whole. The three gunas are three strings of the same single rope. For example, if you have an imbalance in tamas, it will affect rajas and sattva. Each guna is interconnected, so if you modulate the inputs coming into any guna, the other gunas will modulate themselves too, and come to a new balance of order.

This intricate blending is aptly described by Alain Danielou:[211]

Sattva in sattva, consciousness within consciousness, is the nature of Self, Atman.

Rajas in sattva, experience within consciousness, is the nature of the living being, Jiva.

Sattva in rajas, consciousness within existence, forms the inner faculties of manas-mind, buddhi or discerning mind, citta or intelligence and ahankara or ego.

Rajas in rajas, existence within existence, forms the life energies or pranas.

Tamas in rajas, experience within existence, forms the senses.

Sattva in tamas, existence within experience, gives rise to the elements.

Tamas in tamas, experience within experience, forms the inanimate world.

When each guna is in balance with the others, crystallisation of the individual self can occur: individuation. For example, when Rajas action and fire is focused and directed with a sattvic goal it can become a fuel for serving others and *tapasya*, the one pointed fire of deep meditation; rajas is the butter that then becomes refined into the ghee of sattva.

Tamas grounds, roots and centres yourself *and* your expressions. When tamas is harnessed, it becomes a foundation for your creations and soul purpose to manifest in the world. It helps your body-mind stabilize, and when guided it can become orientated to the central channel of the sattvic Sushumna light in your spine. One can then recognise one's limited self and creations, and perform the actions to dissolve these self-created forms.

Transcending the Gunas

The highest level of dispassion leading to Self-Realisation takes place when you are free from all forms of thirst and attachment, including the desires resulting from the inter-play of sattvic, rajasic and tamasic forces of nature.
– Patanjali's Yoga Sutras 1:16

Sattva correlates to the Buddhic mind, the clear, sharp, awake mind that discerns the movements and choices leading to peace, happiness and freedom. Sattva and tamas can both be sources of liking and disliking, attachment and aversion, and these forces act on our minds every day. Sattva likes virtues and is attracted to them, and when it is imbalanced it can make us dislike, and be averse to, that which we judge to be negative or bad – tamas. Indeed, it makes the mind tamasic to do this action!

When we consider that what is good for us today may be bad for us tomorrow, and what is bad for us today may be the most beneficial thing for us tomorrow, we realize that staying attached to any guna, no matter how pleasant, pleasurable, powerful and stilling these forces can be, only limits our own growth.

As Mahatma Dattatreya said, "attachment and aversion create the mind." A virtuous mind is still a mind, and can still lead to suffering. The desire for light alone leads to a rejection of our own darkness. Darkness is both our personal shadow *and* the divine darkness or void from where light and creation arises. In unbalanced sattva we may shun, fear and judge both.

Judgment makes "you" different from "me". However, true sattva is equanimity, and when it cognises the grounded character of tamas, and the wisdom of ever moving uncertainty rajas, we can accept that the nature of the mind and experience is ever changing. *"All paths, whether clear or obscure, are of the nature of the Gunas."* All pathways of experience or Prakriti arise from these three forces.

Altogether, the three gunas *ShaSaSar* are sounded in one beat on Shiva's *Damaru* Drum as one class of sound: three sounds in one pulse. When one physically plays the *Damaru* Drum, a cord hits the drum skin, producing the sound. As this cord moves from one side of the drum to the other, other sounds are created in the movement, sometimes up to three sounds, or a tritone.

Mahadevah is not attached to creation or its experiences. As the maker of creation his name, form and function is *Prabhu*, the Master of Creation, the Master of the City of the Body, a Sovereign independent Being. *This, of course, is you.* Even today in India, an honorific way of addressing a man is calling them Prabhu.

Freedom

One guna or another dominates our daily life at different times during the day, and one guna or another will also dominate periods of our lives. Each guna helps us fulfil a role, an identity, an experience, and "they do their best to fulfil all our desires, until the day of Perfect Union with Source arrives. Then they stop."[212] When one is in this awareness, "one is completely separate from the functionings of the three gunas, and we become Lords and Masters of the three Gunas," like Mahadevah.

The personal self identified and ensnared in the motions and experiences of the three gunas acts without this awareness, as it is fixated on and identified with objects and experiences, thus creating karma.

The transcending of rajas guna occurs when the self ceases identifying with the energy between subject and object, between "me" and "you", between you and another object. This ceasing of need and identification with the energy between two "things" happens when one Recognises that there is only One awareness that is the same within both, and that the awareness within you is the very same awareness within all beings.

Both appear within this same awareness. Awareness does not appear in you: you arise from it. In this, the energy of rajas guna making karmic activities ceases to rule your consciousness. All independent action separate from the awareness within all beings, the awareness within which all arises, ends. This energy is no longer acted upon by you. With this ceasing of self-defined, self-created, self-propelled and selfish actions that create karma and its fruits, comes the end of karma.

As this rajasic doer dissolves into and surrenders to the awareness of the "divine will" and matrix of creation, awareness now moves all actions through you. You have no identity or attachment to what "you" are doing as your causal and karmic body have dissolved.

In transcending Sattva through the vehicle of Sambhu, one sees the same formless living light of Purusha is in you, I, and all beings. This is the highest Namaste. One recognises the goodness in each and every being as also being your own, and your own goodness as existing in every being. You See yourself in everything.

This is joyful, heart-affirming and a wonderful way to exist: in a world of light, made from light, where everyone is an equal part of this light. All you have to do to access this is ask to see this light in others, be it work colleagues, partner, friends and strangers on the street.

We transcend the pull of tamas by entering the void as *"only those who have their minds fixed on the Absolute are untouched by the three gunas."*[213] This Absolute is *nirguna* – without gunas, without form. The way to be free of the Gunas is by continually resonating with, and returning to this absolute awareness, "the impressionless impression", as found in the next chapter.

Sounding the Three Gunas: Sha Sa Sar

The three sounds of the three gunas are known as *usman*, which implies heat, friction and movement.

1. RAJAS-SHA

Mouth open, top of the tongue just touching the top of the lower teeth, medium breath.

2. SATTVA-SA

Cerebral sound, mouth open, tip of tongue touching roof of upper palate.

3. TAMAS-SAr

Mouth open, dental sound, tongue just touching the lower edge of the upper teeth with a subtle "r" sound to end.

Sound Healing Practice

1. Sound *SHA SA SAR* above your head, echoing the serpent Sheshnag. Say this aloud x54 then silently x54.
2. Sound *SHA* into your manipura chakra – navel, aloud x54 then silently x54 until fully connected.
3. Sound *SAr* into your muladhara chakra – spinal base, aloud x54 then silently x54 until fully connected.
4. Sound *SA* into your anahata chakra – heart, aloud x54 then silently x54 until fully connected.
   Slightly contract these chakras physically when you make each sound.

The 13 Dimensions

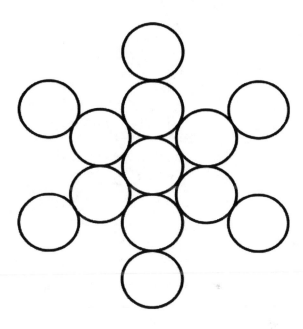

The 13 Classes of Sound that form creation can be seen geometrically as 12+1 circles of creation. The 13 comprise 7+6: 7 circles in the *Seed of Life Geometry* encompassing creation from the quantum field *AIUn* to the completion of the creation of the five elements *LAn*:

The Seed of Life Unfolding

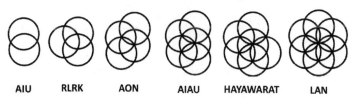

| AIU | RLRK | AON | AIAU | HAYAWARAT | LAN |

and now 6 more in the subsequent unfolding of creation into 13 classes of sound: the *Fruit of Life Geometry.*

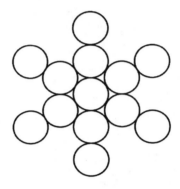

13 is the completion of a cycle of creation. The 13 Classes of Sound unfold to form the 13 spheres of the Seed of Life and Fruit of Life. These are 13 dimensions with 13 different wave-fields, 13 different sets of laws and structures and 13 different containers for consciousness to operate in. In effect, these are 13 different worlds.

If we join the centres of these 13 spheres (spheres are feminine) with straight interconnecting lines (straight lines are masculine) we see the cuboctahedron geometry. This is the only shape in the universe where *the inside is the same as the outside* because its 12 inside radials are the same as its 24 external edges.

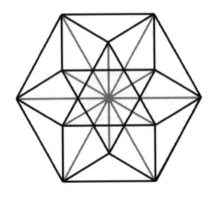

When you make the two into one, and when you make the inside as the outside, the upper as the lower, when you make male and female into a single one, so that the male shall not be male, and the female shall not be female, then you will enter the kingdom.[214]

This cuboctahedron allows the fractal mathematics of infinity to operate in the Fruit of Life, the blueprint of molecular structure and the five platonic solids or basic shapes of material creation.

13 is the number of a natural matrix, a prime number divisible only by itself, hence it is a complete number, totality attained. In music, 12 notes+1=13, the number of return where the 13th note is the repetition of the 1st note, just at a higher octave. This pattern goes on indefinitely. Similarly, the 13th hour in our clocks is the first hour repeated, 12+1. Continuing this theme, there were 12 key disciples around "the One" of Jesus, and Islam has 12 Imams and 1 Prophet, equalling 13.

Time is measured in 13 lunar cycles across the world, from the Mayan Calendar of 13 days in a week and its 52-year cycles divided into 4x13 year cycles. 13 are the number of skies for the Aztecs, and the Sumerians used a zodiac of 13 constellations. There are 12 constellations in the modern Zodiac, BUT there used to be 13, and still are, with the "missing" or "edited out" 13th constellation Ophiuchus being the constellation closest to the black hole in the centre of our galaxy.

In Vedic astrology, the sky or ecliptic is divided into 13 degree slices. The Moon travels 13 degrees across the sky each day, and there are 13 weeks in each of the 4 seasons, as well as 13 weeks between equinoxes and solstices.

This timing mechanism is found in the human body, with 13 lunar menstrual cycles a year for women, and 13 key joints in the body that the Mayans used to calibrate our body with these natural cycles of time. The relationship between the volume of the Earth and the Sun is 13 times the power of 10, or 1/13.01 x 10E5.

There are 13 chakras, not just 7. 13 is "the manifestation of the good or bad generating power", according to famed Egyptologist Schwaller de Lubicz. Thirteen is an uncertain number, a transition: one never knows how it might work. It might be lucky at times and might work against you another time. 13 is a transition number passing from one cycle to another new cycle, and in this sense allows expansion into greater cycles.

The thirteenth Tarot card marks a transformation, the end or death of something, a metamorphosis and a renewal, bringing the beginning of a new era or rebirth. 13 is a number of upheaval so that new ground can be broken. The number 13 has great power. If this power is used for selfish purposes, it will bring destruction and dis-ease. If it used for elevatory purposes it will bring growth. Power is neutral: it just depends how you use it.

13 is seen in both the forming of life in the womb, and in death. In biological science, it takes 12 male sperm surrounding the 1 feminine egg in order to fertilize and impregnate it, thus creating the foetus. In India, when a person dies, the 13th day marks the point when the soul's *koshas*, or bodies, dissipate. Ceremonies to help the soul pass through the different dimensions (Bardo) one encounters upon leaving the body are done for 13+1 days. The 14th day is the final exit of the soul from the third dimension, as we shall see in the next chapter.

In quantum physics, there are 13 superstring dimensions, with the Universe governed by thirteen fundamental constants. As R. Allendy shares, "13 produces one cycle of perpetually identical renewal (1+3=4). The mechanism of this organization submits the Universe to a permanent mode of oscillation."

The Triple Spiral is seen as a feminine symbol of creation across the world.

13 is a key number in all sacred traditions worldwide and is a feminine cyclic number. In the Gnostic Gospels and *Pistis Sophia*, Christ shares that there are 13 heavens or *Aeons*. The soul falls from the 13th Aeon, a profound yet not divine position, to the material planes when born. The 14th heaven is seen as divine, whereas everything below the 14th is ruled by matter.

As Christ shares, "beyond the 13th Aeon God is unnameable. He has no human form; for whoever has human form is the creation of another.[215] He has a semblance of his own – not like what you have seen and received... that surpasses all things and is better than the universe. It looks to every side and sees itself from itself. Since it is infinite, he is incomprehensible. He is imperishability blessed."

What does this mean? In the next chapter, we shall see...

HAL: The Supreme

तत्त्वातीत: पर: साक्षी सर्वानुग्रहविग्रह: ।
अहमात्मा परो हल् स्यादिति शम्भुस्तिरोदधे ॥२७॥

tattvātītaḥ paraḥ sākśhī sarvānugrahavigrahaḥ
ahamātmā parō hal syāditi śambhustirōdadhē
27

All elements of creating come into form
when Seen by the Supreme Witness
in Its light of loving-kindness

It Sees goodness in all beings,
and Sees Itself in everything

Seeing the creation was Good
in the supreme detachment of pure love
transcending all elements of creating,
Sambhu declared: *I AM the Soul Supreme HAL*

Sambhu, Source of Bliss,
with all creation manifested, embodied and fulfilled,
Vanished away.

Maheshwara-Sambhu has now completed creating a new universe and a new human being called *Aham* through the 14 Classes of Sound.[216] The ingredients for this creation are the *Tattvas*, which literally means "that thou", meaning the elements you are made of, "that thou art". Your body-mind-spirit, all experiences, all forms and the entire universe are made of these elements or tattvas, all of which have now been created and completed.

You and the universe have now all been sounded into embodiment in the previous 13 classes of sound, and are now fully manifested by being "seen" by the Supreme Witness *parah sakshi*. This echoes quantum physics, in that the observer influences and co-creates whatever is being observed.

The witness or *sakshi* first appeared in the third movement of creating *EOn*, where the witness "Sees" creation into manifestation by observing it in conjunction with Maya, the architect of creation. Now, the witness appears for the second time, but is named differently as the supreme transcendental witness *Parah sakshi*, beyond any vibration and beyond Maya.

All parts of the creating process are sandwiched between these two faces of the Witness: in the beginning of creation *RLRK-EOn*, and now the last act of the creative process *HAl*. Everything is Seen and observed into existence.

The Knower or supreme transcendental witness *Parah sakshi* has no attachment to any experience, process, movement, or any part of creation: it is just peace, bliss and Presence. What can be known, which is all experiences, thoughts, processes and objects, is limited. The Knower of this known, the supreme transcendental witness *Parah sakshi*, is unlimited.

Parah sakshi Sees all creation as good, Seeing all beings and all creation through the eyes of loving kindness, Seeing all that has been created in a shower of blessings, blessing all that has been created and establishing all creations, all its children, in love.

Parah sakshi loves creation into being. Indeed, all of creation knows it is loved now, and this action seals the creating process into full embodiment and manifestation. This Seeing instils within all sentient beings a deeply felt love for living, a love for life, an inner knowing in one's essence that it is good to be alive. As *Parah sakshi* loves creation into existence, It establishes love as our essence and our very reason to be alive. This is how we were created.

We can feel this as gratitude, an inner thankfulness that we are alive, and in our deep heartfelt desires, so we can enjoy this precious opportunity to live and to love. Whenever we feel these feelings, we align to the blueprint of creation.

This Seeing of *Parah sakshi* arises from bliss and silence. From silence arises the sound of every form, and the form of every sound. All sounds are included in silence because all sounds arise from silence; in silence, one can know all creation, and all the sounds that compose creation.

All sounds ultimately lead into silence, and from silence all sounds can be known. This is the ultimate purpose of sound for us: to bring us into the silence of our origin. In the Sutra, this is accessed through the doorways of *AH* at the start and *HAl* at the end.

In the beginning of creation *AlUn* in Itself and through Itself appears in a wave of clear light, tremendous energy-desire and loving joy. The witness then Sees Itself looking at Itself in the first reflection of Itself: Maya.

The witness Sees the egg (EOn) that gestates this reflection, dividing and multiplying in more reflections to bring forth more expressions of this ecstatic dance. The witness observes the first time he and his love see each other, shimmering shivers of desire and delight as they sing worlds into being.

The witness then Sees a vortex of five iridescent elemental streams and their geometries coming forth, swirling in seas of resonance, moulding and shaping matter to cloak Itself with.

The witness Sees a mirror of itself in the human form, seeing the ability to make these sounds of creation through human speech and understanding, revealing a way of differentiating inside from outside as another play and pleasure. The supreme witness Sees how each cell moves, releases and multiplies itself in order to become matter and continue this dance.

The witness Sees itself in sound, taste, smell, touch and form, bathing itself in sensorial delight. The witness Sees the weaving of five pranic currents dancing in vibrational flows, forming the trinity of the mind in the human foetus. The witness Sees its living light animating and forming every being.

The witness witnesses all comings and goings, all formings and all disappearings, all sensations and all thought forms, all time. As the silent watcher of the dancing pairs of opposites, the witness watches in love as the pairs create the ground through which the dance is possible.

The witness Sees three spirals of light, movement and stability in every wave. Then the Supreme Witness beyond creation, in the supreme detachment of pure love, leaves the creation, vanishing in the bliss of Being.

First one sees the Self as objects, then one sees the Self as void, then one sees the Self as Self, and only in this last is there no seeing, because seeing is Being.
– Ramana Maharshi

The Blessing

Sarvānugraha vigrahaḥ is a beautiful phrase, pure poetry in sound, a sonic bath in the heart. *Parasakshi* the Supreme Witness, in the supreme detachment of pure love, Sees goodness and love in all things and in all beings, Seeing Itself in everything, in all forms, in all life.

This awareness expresses itself as everything and everyone. This awareness appears as everything and everyone. There are no others, nothing other than itself. This absence of otherness is love.

In this, *Parasakshi* loves creation into being, and all of creation knows it is loved now. When we open to receive this Grace, we see love in all forms, in all life, and in our own self.[217] This love is our true Self.

When the One reveals Himself as a form, this vigrahah exists eternally. He is form vigraha, but at the same time He is not. His Presence reveals itself in everything; one realizes it is He alone who appears as Being and also as becoming. A seeker... comes to realize that his Beloved resides in every creature, and every creature in Him, echoed in the saying Devo bhutva devam yajet – only by becoming identified with the Lord can one worship him.
– Sri Anandamayi Ma

Sambhu

Shiva is *Sambhu*, an eternally happy, blissful, joyful and generous auspicious goodness. We can have tastes of the uplifting joy and happiness of Sambhu in song, mantra, prayer and devotion, but this infinite awareness is not limited to any experience, identity or form. This infinite awareness is where we truly find Sambhu, as only the infinite can know the infinite.

In its peak experience Sambhu is bliss, the One who enjoys all bliss ceaselessly, the drunken bliss of *samadhi*. Samadhi occurs when one's breath and heartbeat slows down to a few beats per minute. The "I" dissolves in potent ecstasy as the pineal gland opens, flooding the brain and body with bliss hormones. Once this is experienced, one Realizes this is the home one has always been looking for, and one never wants to leave this nirvana.

In samadhi, memories, karmas, *samskaras* or seeds of causal suffering, *vasanas* or patterns and tendencies, can dissolve forever. But despite our best efforts, aspirations and yearnings, samadhi only happens through Grace, *sarvanugraha vigraha*.

This is the most abundantly profound gift and act of kindness a human can ever receive.

Ahamatmo Paro Hal

Behind the apparent multiplicity and diversity of creation, there is a single reality or awareness which cannot be named or defined, as names and definitions refer to single parts. Awareness has no name or definition. In this single awareness, there is just Itself. Everything is an appearance of this awareness, of this single, infinite indivisible whole.

Everything arises in this Awareness, "the essential irreducible element of ourself, the only aspect that never appears or disappears. It is never changed by anything that it is aware of. Just as one who watches a movie is not implicated by anything taking place in the movie, awareness witnesses all experiences and is utterly intimately one with it, and at the same time independent of it.

Awareness is never moved or changed, hurt or harmed by anything that takes place, and it is peace, prior to and independent of the contents of experience, a peace independent of what does, or does not, take place in experience."[218]

Sambhu Sees this creation as good and complete in his declaration *Ahamatma Paro Hal: I AM the Great Soul Beyond*. The only thing Awareness can express is this I AM. Awareness by itself cannot perceive anything because only the infinite can know the infinite. The infinite cannot perceive the finite directly as there is no object-subject in the infinite. If the infinite is to perceive anything finite, it must do so through the finite mind.

To perceive the finite activity of the universe, and to create finite universes, Awareness must do so of its Will (mentioned several times in the Sutra), and create a finite mind and form in order to perceive this activity.

I AM is a gateway to Awareness *and* to the manifestation of creation *depending on which way you enter it*. I AM stands in between Awareness and the finite mind. If a finite mind passes through I AM into awareness, it loses its limitations and becomes this awareness. If awareness passes through I AM in the other direction, it apparently gains a quality and seems to become finite. (These two directional portals lie at the beginning and end of creation in *AIUn-HAL*.)

This is why I AM is said by Krishnamurti to be "the first and last freedom." It is the portal through which awareness passes in one direction, and apparently loses its freedom, and the portal through which finite mind passes in the other direction to gain its freedom.

In the Sutra, *Aham* I AM is the body of the universe in all its forms, generated from the first sound *A* to the last sound *HAl*. *HAl* is a gateway out of this universe of processes, experiences and forms. In *HAl*, the whole universe is viewed through a singularity point or "eye" of love.

H is the 16th and last Sanskrit vowel sound, denoting consciousness being in a singularity point, where an internal creative force *and* an external creative force are simultaneously occurring.[219] Here, if you observe in one way, one finds that nothing has been created. This is the internal creative force. If you observe in another way, one finds that everything has been created. This is the external creative force. So, by observing in one way, nothing is created, and if at the same time you observe in another way, you will find that everything is created.[220]

The Supreme Witness *parasakshi* Sees nothing has been created, as the infinite can only perceive the infinite; *and* It also Sees everything has been created in love, as part of Itself. This is its Game of freedom, delight and love.

This supreme detachment of unconditional love is empty, has no attachments, no needs and no identity to any form, being, process or sound. Unconditional love simply Sees Itself, as love, in each being. Creation is full of love *and* is nothing. This *Param Siva* is everywhere, that is why it is nowhere. It is all levels, and therefore has no level. The One Being who is everywhere is nowhere. Hence, It disappears at the end of the Sutra.

The Fourth Face of Shiva

The birth of the world, its maintenance, its destruction, the soul's obscuration and the soul's liberation, are the five acts of His Dance.
– Raurava Agama

Dance expresses and transforms energy. No energy is ever destroyed or lost in a dance: it just changes from one form, state and wave to another. In *nritya-murti*, Nataraja Raja dances the five faces of Shiva, the five universal powers of creation, sustenance, destruction, veiling and liberation.

The face of Destruction is revealed in the first verse of the Maheshwara Sutra by Nataraja Raja dancing in *samhara*, where forms and creations are destroyed. This is seen in the fire in his upper left hand held in *ardha-chandra* or half-moon *mudra*.

The face of Creating *Srishti* is shown in his upper right hand beating the drum from which the 14 classes of Sound emanate. Shiva's *Sthiti*, Sustaining Creation, has his lower right hand held out in the blessing of *abhaya mudra*, "be fearless, fear not."

The name of Shiva as Maheshwara is his fourth face, masking and cloaking, known as *tirodhana shakti*. This face hides, veils, obscures and cloaks pure awareness. The Maheshwara Sutra is the revealing of Shiva's fourth face,[221] showing the creation of the finite in order for the infinite to perceive the activity of the universe.

The Sutra shows the steps of the blueprint of creation, from which all possible experiences can arise. Each step of creation is like a layer of clothing we put on, and can take off. When we no longer identify with any of these layers, experiences and processes, then that which is ever Present remains: pure awareness.

This is Shiva's Fifth Face of *anugraha Shakti,* or Freedom/ Liberation through Surrender. All awakened beings throughout history who tried so hard, meditated so deeply and sought for wisdom so assiduously, gave up their search as part of the process of anugraha, *of surrender* and grace.

The full flowering of anugraha is a destiny no one knows the timing of: everyone has his or her time for awakening. The mystery of anugraha is that it comes when it comes, like "a thief in the night" delivering total transformation.

The Maheshwara Sutra is Shiva's Tantra – being fully engaged in and fully enjoying creation, whilst being totally unattached to any part of it in the supreme detachment of pure love. In order to fully engage in anything *is* to be fully detached at the same time. To be fully present is to be fully absent.

Black Hole White Hole

The Vedic *pralaya* states that an endless succession of universes have been created, are being created, and will be created, and an endless succession of universes have been dissolved, are being dissolved, and will be dissolved, to then be endlessly recreated ad infinitum, one cycle flowing into the other.

The "big bang", or more accurately "silent clear light" that marks the creation of a universe, arises after the dissolution of the previous universe, as it completes its function and purpose. (This function differs according to the purpose of why the universe has been created.) As this dissolution occurs, the universe reduces itself into a concentrated, miniscule and incredibly dense mass, a Singularity or *Bindu* point.[222]

In this understanding, the universe we live in is one of many universes that exist, and there are countless other universes that have existed, and will potentially exist in the future. Each universe is modelled on different laws: one universe could have gravity and the elements, another could not. Each universe has different forms of life, different laws of physics and different expressions of time and space, with some having no time or space as we know it.

The Maheshwara Sutra shows how waves of vibration starting from *AIUn* create all possible forms and experiences, unfolding the many functions of the universal creative process. These waves then dissolve through *HAl*, leaving the creation just created, *to then reappear again* as waves of vibration in *A I Un*, to then dissolve again in *HAl*, and so on, cyclically repeating itself in the transformation from form into formless into form.

All the forms of the universe occur between these two voids of *A I Un*, the beginning of creation expelling itself via the white hole of Shiva's *Damaru* Drum, and the other void or black hole that sucks in all creation in the great disappearing act of *HAL*. Creation lies between these two voids *which are the same but different*. The back hole void *HAL* is not the same as the first void or white hole from where *AIUn* emerges.[223] The words and terminology (the supreme witness *parasakshi HAL* of the black hole and the witness *sakshi* of the white hole *AIUn*) used in the Sutra to denote this are different.

THE SECRET OF LIGHT

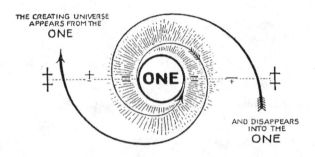

A black hole and white hole are two sides of the same coin, as you enter a white hole through a black hole. A black hole is a singularity, an indivisible point of zero volume and infinite mass, sucking in all matter, all forms, all experiences and all light. This is the "end" of a form, the end of its timeline for existence, and the end of time and space for that form, all of which corresponds to HAL.

The white hole *AIUn* or "quantum dot" is the *Nada Bindu*, the concentrated point from where all sound and creation begins. A white hole arises from the "other side" of a black hole and could not exist without it. And vice versa: a black hole arises from the other side of a white hole. Everything in creation works in mated pairs, perfected polarities (not duality) as seen throughout the Sutra.

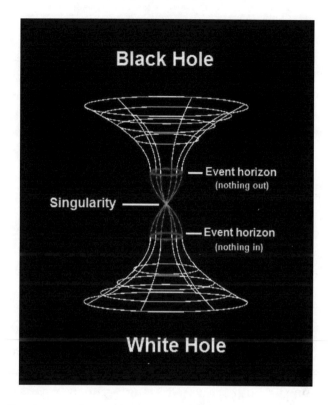

This diagram can be seen as the hourglass shape of Shiva's Drum.

There are infinite amounts of matter and light a black hole can suck in, which then emerges in a different state, transformed, through a white hole. One could say whatever goes into a black hole is dissolved, to then be resurrected anew through the white hole in a different way.

Interestingly, a black hole can *only* be entered from the outside, and a white hole cannot be entered from the outside. This means that *the interior of a white hole is cut off from the universe's past and future*, which means a white hole is the eternal present moment, the quantum field, as found in *A I Un*.

This eternal present moment is where love, freedom and creation occur, in *sarvanugraha vigraha,* and this is the goal of all spiritual traditions: to live in the present moment, to be present in the power of Now.

Creation, and ourselves, arise from a white hole and dissolve into a black hole. We come from the void and go back to the void: in between, there is this life, a passage (as seen in the hourglass drum of Shiva) between creation and dissolution, a temporary phenomenon that is transient, that comes and goes.

From nothing into nothing is the journey of creation. There is only one awareness that lasts, that is eternal: the rest passes away. Hence most spiritual paths focus on the eternal, that which never changes, that which never goes away. Only the awareness which is beyond the phenomenon of energy and vibration (Maya) remains, and this is found within your own Self. Everything else will vanish.[224]

Our universe will end, to be born again in a new way, with a new purpose, a new set of laws and parameters. This potentially never-ending process of creation and dissolution, of our universe being birthed from a white hole via the end of the previous universe through a black hole, continues ad infinitum. We end all identification to these processes through *HAL*.

Orion: Shiva's Damaru Drum

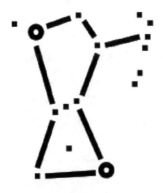

Bindu

NavaYoni Yantra
9 Cosmic Wombs

Shiva's Drum
Rhythm of Creation

The shape of Shiva's *Damaru* Drum is remarkably similar to Orion. Ancient Tamil texts further corroborate that Shiva comes from Orion. In this context, the outflowing of creation *A I Un* arises from the Orion white hole, weaves the matrix of creation, and then dissolves into the black hole of *HAL*, which presumably lies at the centre of our Galaxy near the constellation Sagittarius.

The Five Types of Black Hole

In the Maheshwara Sutra all creation lies between two voids: we arise from a white hole and disappear into a black hole. We are created, travel on the spiral cycle of incarnation marked out by the 42 sounds of consciousness, and are recycled through the black hole to return again into incarnation.

For "one's" separate individual consciousness to dissolve into a black hole and NOT re-emerge into the cycles of reincarnation, i.e. to dissolve one's separate self into pure awareness, requires two things: one is to focus on this pure awareness beyond everything else. The second path, which is the most common on earth, is to cognise and release all our emotional, mental and spiritual conditioning, traumas and beliefs that prevent us from living as this awareness.

This includes all fragmented or wounded aspects, disempowered aspects, aspects absent of love and aspects that are unconscious and have not been seen, cognised and integrated. We have to have awareness of ourself through direct experience.

If we have not healed and developed these aspects, we get recycled into creation again and again, until we learn to develop these aspects. If we immerse into a black hole totally, i.e. we immerse into pure awareness beyond our ideas of a separate self, we travel into the eternal present moment and pure creative power of a white hole, which then resurrects one into Self Realization.

To surrender to this infinite immense power and love requires one has no fear and has mastered their self. There can be no wounds hidden, no regrets and total openness on all levels. This love is incredibly powerful, a love that brings one to their knees, a love that dissolves who you think you are and what the world is.

To enter a black hole and survive *you do not enter straight in*. One can only enter a black hole successfully through the correct angle of a spiral, and this angle is calculated through Phi – the Golden Mean Ratio Spiral in the 14/42 spiral vortex of the Maheshwara Sutra.

Many initiates in ancient Egypt used to do this journey by utilizing the initiation chamber of the Great Pyramid with these techniques.[225] In today's world these techniques are secret, and regardless, this Age is not designed for these past techniques. We live in a different frequency waveband now, and our initiations are different, yet similar, in many ways.

To access a black hole one needs to be in a singular one pointed focus or Bindu point, where all aspects of your consciousness, mind and soul converge. In this concentration of all forces of thought, attention, emotion and presence, you go through the

Bindu point into the "no point" of infinity, a state characterised by deep bliss and the dissolution of yourself.

A black hole is a singularity, so one-minded that it only ever does one thing; dissolve everything totally. If "one's" consciousness is to move through this, you have to totally embrace being no-body and no-thing. In experiencing such an immersion, any aspects within you that are not in flow with this unbroken continuum of love and consciousness will arise.

The only way one can enter a black hole is through a pure, open, deep love. This love is formless-living light. The love in the core of the *anahata* heart chakra is unstruck, unformed, formless love, and it is only this love that can take one through the heart of a black hole and then into the white hole to be resurrected into eternity in form.

No mind, no matter how brilliant, how accomplished, how sure of its own clarity and light, can enter a black hole. All mind and all matter dissolve in a black hole, so one could say the whole universe is mind matter, as the formless is always and only peace, presence, bliss and love.

As Sambhu declares in the last verse of the Sutra, it is only through the heart of grace and loving kindness, and by feeling the love in all creation, in all beings, in yourself, that one can disappear into the black hole, to then emerge through the white hole as the infinite in form.

The white hole re-creates and resurrects the being that has just ended in the black hole, into an immortal being. Throughout history many beings in many traditions have followed this pathway, from Shiva to Christ to Osiris and more, as we shall find out in the next chapters.

There are five types of black hole covered by five spiralling veils which constantly whirl in the consciousness of the separate self. These five hold keys to the transformation of your body, emotions, mind, perceptions and self-consciousness. These five veils are made of everything you have ever, in all time, somatically felt, experienced, thought, believed, identified with, perceived and therefore constructed yourself to be.

As one's separate consciousness enters the spin of a black hole, you, as you feel, think and know yourself to be, begin to dissolve. We may experience this when we are with enlightened beings, when doing spiritual practices, experiencing true Tantra, experiencing psychedelic ceremony or *Soma*, and in moments of grace.

In the spin of a black hole, you die a little in each revolution you make around its centre. In each revolution you make, you vibrate faster and faster, so there is less and less of the separate self or *you* present. The movie of your life dissolves. You eventually vibrate so fast and your frequency becomes so high, that you become still. In this stillness, you enter the heart of the black hole awareness.

The five types of black hole are: an uncharged, non-rotating black hole; an uncharged, rotating black hole; a charged, non-rotating black hole; a rotating, charged black hole; AND a pure, rotating spinning force.

The uncharged, non-rotating black hole is a portal to the transformation of your DNA, which mutates to bring one into neurobiological Self Realization. The uncharged rotating black hole is a portal to the transformation of perceptions, enabling you to See, Witness and Know anything.

The charged, non-rotating black hole is a portal to the dissolving of your mindset and your mental constructs. Your mind then becomes the *servant* of awareness, and the mind is recognised as appearing in awareness rather than awareness appearing within it.

The rotating, charged black hole is a portal to your emotions and feelings, where all emotions are embraced and therefore transcended, allowing you to enjoy them in a different way just as Shiva and Shakti do in their dance of Self Delight that creates the universe. Your emotions become feelings, and your feeling range or *rasa* broadens into infinity, without limits.

The last type of black hole is a pure, rotating spinning force: the transformation of the separate self in the Dance of Shiva. The faster the black hole spins, the more mass is lost by any object entering it, until there is nothing left of the object. This mass *is the sense of yourself,* as the Sutra has shared in detail.

To enter a black hole, let go of yourself and surrender to love: then you will emerge resurrected.[226]

In entering a black hole, you can realize the power of creation in its other side of a white hole. Harness these powers and you can build universes, for a black hole/white hole enables all the notes of matter to play, to form, and to dissolve.

Fourteen: The Resurrection Code

The universe and the human being are composed of 14 Classes of Sound. There are also 14 Names of God in the Maheshwara Sutra, all of which can be used together in one stream of sound meditation, *as well as* separately in individual mantras.

1. MAHESHWARA – Maha-Ishwara, The Lord of All, in All
2. NATARAJA RAJA – King of Kings, Lord of the Dance
3. SIVA – The Auspicious One
4. SAKTI – Feminine Creative Energy
5. VISNU – All-pervading upholding consciousness
6. ISHWARA – Lord of Creation in creation
7. VYAPAKA MAHESHVARA – All-pervading existence
8. MAYESHWARA – Maya and Isvara in Union
9. BRAHMA – The creator of forms and space-time
10. VIRĀDRŪPAṀ – Universal Form
11. PARAMATMANAH – Supreme Soul
12. PURUSHAM – Primordial Lord
13. MAHĀDĒVAḤ – Great God beyond all gods
14. SAMBHU – Lord of Bliss

It is important to understand that these 14 Names are part of your vibration, your essence. In one sense they are your Names, your qualities, for you are part of God. These 14 Names are portals for you to enter/experience This as reminders of your true nature and creative power. The universe is contained within you.

In the *vartika* or commentary Upamanyu wrote on the Maheshwara Sutra called *Tattvavimarśinī*, in *v. 13* about *EOn*, it is said that Shiva created 14 worlds and 14 chakras from himself: "He Himself became all that."

This is echoed in the *Brahmāṇḍa-Purāṇa* 1.2.20, where the Cosmic Egg *EOn* consists of 14 *lokas*, worlds or dimensions: *Satya-Loka, Tapo-Loka, Jnana-Loka, Mahar-Loka, Svar-Loka, Bhuvar-Loka, Bhu-Loka,* and the underworlds of Atala, Vitala, Sutala, Talātala, Rasātala, Mahātala, Pātāla. These underworlds are chakras below the muladhara root chakra in the legs, knees and feet.

These 14 chakras descend from *Satya Loka* or the soul star chakra ending at *Patala* or the underworld of the foot chakras. Ways to work with these chakras are part of the Saivite tradition. Curiously, it is also said in Tibetan Buddhism that there are 14 different star systems where *Dzogchen* meditation is practised.

The number 14 is further echoed by Upamanyu in *Tattvavimarśinī*, who shares that all tattvas or elements of creation ranging from Shiva to Prakriti are created through the 14 vowels *a, ā, i, ī, u, ū, ṛ, ḷ, ē, ai, ō, au, aṁ, aḥ.*

Fourteen also relates to the 14 enclosures of the *Sri Yantra* or *śrichakra* (master of the wheel or master of the chakras). The Sri

Yantra is the most famous of all Yantras, the map of all Creation in one geometric form. All the tattvas or elements of creation are found in the Sri Yantra.

The Sri Yantra also shows the union of male and female, Shiva and Shakti, in its interconnecting triangles. This of course is also the Dance of the Maheshwara Sutra, where the mated pairs that are Shiva and Shakti in their different forms and names create the universe.

The 14 enclosures and 42 pyramidal points of the Sri Yantra are reflected in the 14 Classes of Sound and 42 syllables of the Maheshwara Sutra. These points mark the incoming and outgoing waves of creation of the toroidal spiral.

As a sacred geometric map, the Sri Yantra depicts sound, as all geometry is vibration, is sound. The 42 sounds of the Sutra activate this sonic geometric map in your consciousness when one meditates on the Sri Yantra in a specific way whilst making the sounds.

The 14 Parts of Osiris

Fourteen is the number of steps required for transformation and resurrection into pure awareness in three ancient sacred traditions: the Vedic, ancient Egyptian and Gnostic Christian, as told through the stories of Shiva, Osiris and Christ.

The Maheshwara Sutra has remarkable similarities to the transformations that Christ and Osiris undertook in order to awaken. The Sutra contains original wisdom that unites Eastern and Western spiritual traditions, and it contains missing keys to the Western spiritual tradition.

These wisdom processes all hearken back to the original creation story of the Vedic *Purusha*, the Cosmic Man, who "sacrifices" his body parts in order to create the universe. This is paralleled in Shiva creating the universe from himself in the Maheshwara Sutra's 14 classes of sound, and the 14 body parts of Osiris[227] which are cut up and "sacrificed", revealing the initiation process of awakening.

This Osiris story became part of the Christ story later in history (much of Christianity is derived from Egyptian wisdom), where Christ Yeshua sacrificed himself in his crucifixion journey of the 14 Stations of the Cross. Even today Christians eat and drink of the body of Christ in Communion.

Osiris and Christ both took their journeys of transformation into awakening in 14 steps, following the even more ancient initiatic pathway of Maheshwara Shiva. This is no coincidence to have the revered Lords of three major sacred traditions (which are in truth one original teaching separated over time, language and geography) following the same journey into awakening in 14 steps! The history of the Western world *and* its deeper experiential wisdom cannot be fully understood without this wisdom.

The wisdom behind this process is encoded in sound, and the processes of creation it unfolds, creates and describes.

Osiris was one of *the* most important Egyptian gods, and was what every Pharaoh sought to become through the alchemical processes of *Osirification*. Osiris is seen as a King and Lord, a Christ figure.

In his resurrection story, Osiris is murdered by his brother Set. The first version of this story details how Set kidnaps, kills and entombs Osiris in a sarcophagus, and then cuts up and hides his body in the desert.

This sarcophagus is tailor-made for him, meaning that its dimensions were made to encompass the blueprint Osiris embodied. His corpse is discovered by his wife Isis and sister Nephthys in the form of birds, who resurrect him by the beating of their wings[228] in a powerful vibrational hum.

This revivification of Osiris happens in the sky-earth Goddess Nut's womb/sarcophagus/birthing chamber,[229] a cellular re-generator. The 14 pieces of his body are gathered (*jnq*), amassed (*s3q*), united (*dmdj*), drawn together (*shn*) before being joined (*m'b*) and connected (*tjs*) to each other. *The conception of Osiris' son Horus takes place at this moment.*

The second resurrection story of Osiris is when Set finds his corpse in the desert and cuts it into 14 pieces, throwing each portion into the Nile, the River of Life. Osiris' freshly-born son Horus searches for each piece, and succeeds in collecting all 14 parts which are then put into Nut's sarcophagus or birthing chamber to be reconstituted, regenerated and recomposed,[230] resurrecting Osiris.

His 14 body parts are never mentioned specifically, as in head, arms, legs etc. BUT the 15th part that resurrects Osiris is specifically mentioned as being his *lingam*, his generative organ and creative power, which is "given" back to him through the magic power of his wife Isis. (This is similar to Christ, as the first person to see him in his tomb after his resurrection was his wife Mary Magdalene.)

The lingam is connected to earth element, to form, regeneration and life. In the Pyramid Texts, Osiris is re-born when the earth, which covered his corpse, is removed. Fourteen is the number of Osiris without his lingam, and the *number for Osiris without his feminine complement.*

The 15th step of Osiris having his lingam, his generative power, is accessed through 14, the gateway out of space-time. Yet it is only through his feminine creator complement Shakti/Isis that his process of resurrection is completed through her "kiss of life".[231]

Osiris is re-born into *Atum*, the first Creator, a Maheshwara figure or Great God. Osiris is not born again into his previous state of consciousness, but into an exalted, ascended consciousness, just as Christ was in his resurrection, and as Sambu points towards in the Maheshwara Sutra.

Every Pharoah in Egypt tried to follow this initiatory sequence in some way for thousands of years after these events, for every Pharaoh wished to become Osiris, to become immortal and live forever.

The Pharaohs deified Osiris, and created 14 specific star shafts in the King's Chamber of the Great Pyramid that connected to 14 stars, especially the Orion Constellation. These 14 star shafts were used as part of the Osirification process for initiates to have the direct experience of becoming at-one with all of creation, in pure awareness.

The Forty-Two Notes of Ma'at

Do you not know that Egypt is an image of heaven... in Egypt all the operations of the powers which rule and work in heaven have been transferred to the earth below... the whole Cosmos dwells in our land as its sanctuary.
– Walter Scott, *Hermetica*[232]

The Egyptian Edfu Temple Texts state that Egypt was created as a celestial kingdom, as above so below: the kingdom in the stars above, particularly Orion-Osiris, was mapped out in the land of Egypt below. This kingdom, both above and below, was composed of the 42 harmonics of Ma'at, the universal principles of order, truth, balance, righteousness, morality, justice and harmony.

These 42 harmonics of Ma'at became the 42 different kingdoms or *nomes* of Egypt, with each kingdom having their own ruler or *nomarch*. Each nomarch was seen as a god/goddess or *Neter*, wielding a specific power and function. Each of the 42 nomes contained a specific state of consciousness, specific laws and *Neters* and a particular defining of the universal creative process.

Each harmonic and each state of consciousness contained specific qualities, which together formed the whole kingdom of heaven on earth. The 42 kingdoms together created the sum total of all its parts: Egypt as a celestial kingdom on earth, a portal to universal consciousness.[233,234]

The 42 were accessed on earth through the spiritual practice of *Ma'at*. Known as *Rtam* in Sanskrit, Ma'at is the universal order that holds creation in balance, through the open heart. In ancient Egypt, the 42 Steps of Ma'at were a spiritual practice that priests and priestesses undertook as part of their purification, healing and enlightenment.

In this self-inquiry practice, each person's heart and deeds were examined, cleansed and then "weighed" on the scales of Ma'at by Anubis, Thoth and 42 judges, who ascertained whether the heart was as pure and light as Ma'at's feather.

If the heart was as light as a feather, one would pass through the Hall of Ma'at into the Kingdom of Osiris and the Boat of Ra to become a liberated soul free of the wheel of reincarnation. In successfully passing through these 42 steps of Ma'at, one reached the completion of their incarnations on earth, and entered eternity. Emotional purity, integrity and knowing of one's Self were the gateways through which the unified heart-soul could fly freely into the heavens as a liberated soul.

Osiris, the Lord of Egypt and god of the underworld, was the presiding judge over this process and the principal *guardian between life and death*, between this world and the next. He was the final arbiter of who was to be resurrected into eternal life on the "Boat of Ra", and who was not ready to enter eternal life, *i.e. who would become reincarnated*. He had the final say in which humans could pass through the portal that death (*Hal*) opens for us to enter the "afterlife", or cosmic consciousness and awareness.

Christ, the Kumaras and the 14 Stations

The Maheshwara Sutra was given explicitly to *Sanaka Kumara*, one of the four Kumaras or Mind Born Sons of Brahma. These four were created directly from Brahma's consciousness, not from a womb or woman, to continue Brahma's work of propagating creation and giving humanity a pathway into God consciousness.

Brahma immaculately conceived and directly begot the Kumaras as *saktyavesa avataras*, or indirect incarnations of God. When God comes directly to earth in form, as with Krishna and Rama, He is called an avatar or *saksat*, and when God empowers a living entity to represent him as an agent of sorts, such a being is called an indirect or *avesa* avatar.

Once they had been created, the Kumaras looked at humanity and the world and were not interested in joining in with what was happening. They did not want to have families, offspring or sex. Yet, they also saw the nature of human suffering, understanding that their reason for being created was to help humans find ways into their source and God consciousness.

As avesa avatars of our planet Earth and humanity, the Kumaras periodically bring us new teachings and embodiments of the divine in order to help guide, inform and elevate us into becoming a conscious part of God Consciousness. They have been doing this for many thousands of years.

As sublime embodiments of God and teachers of awakening, the Kumaras have no equal. In the *Srimad-Bhagavatam* 2:7:5 it is said that, "these four are incarnations of the knowledge of the Supreme Lord, and explained transcendental knowledge so explicitly that all the sages could assimilate this knowledge without difficulty. By following in the footsteps of the four Kumaras, one can see the Supreme Personality of Godhead within oneself."

Kumara means "child" and they are always pictured in Indian literature as looking like five-year-old children, beatific, innocent and full of love, like cherubim angels.[235] They are seen, along with Krishna and Visnu, as being pure love embodiments of the divine.

As it is said, Brahma created the four Kumaras to propagate and help humanity by empowering them to share four streams of wisdom: the science and study of creation; yogic mysticism for liberation of the soul; the art of detachment and witnessing, and meditational practices in order to attain enlightenment. The Kumaras then inaugurated their own spiritual lineage, the *Kumara-sampradaya* or *Nimbaraka-sampradaya*,[236] and spread these teachings in order to share pathways into the source of creation and God Consciousness.

Sanathana Kumara, "the continuity of eternity" is the first and oldest Kumara, and is Known as the energy that flows through all enlightened teachers of advaita or non-duality, which is the end of all knowledge and all seeking.

Sanat Kumara, the "eternal divine creation" is the initiator and first teacher of the alchemical, transformative wisdom path. He is mentioned in many ancient Vedic texts and in the Kashmiri Saivite tradition as being the first initiator of many enlightened human teachers, who then passed on the wisdom and practices of transformation to many others.

This lineage then spread around the world from Asia to Persia, and is most famously known in the Western world through John the Baptist's teachings and lineage, of whom the Knights Templar were one branch. Sanat is also known in esoteric circles as Lord of the World, responsible for the evolution, guiding, elevating and caring for humanity in a detached but fatherly way. He is closely connected to Venus.

Sanaka Kumara the "eternal being" is his brother, a teacher and supporter of humanity as part of the Kumara Avatar. Sanaka is the Kumara mentioned at the beginning of the Maheshwara

Sutra, and is the head "scientist" of the lineage that studies, practices and implements the science of creation, the science of consciousness.

His incarnations throughout history worldwide include Lucifer, Prometheus, Set and Judas. He, like his father Brahma, has the brilliance of creative light, where the greatest light casts the greatest shadow. Sanaka plays the unglamourous, much judged but essential role of instigating and embodying the polarities that fuel the creative processes of our universe, which become dualities when perceived through the fragmented finite mind.

His role throughout history has been to play "the bad guy", the polarity/duality to the "light guy", as seen in his dance as Set with Osiris and in his dance as Lucifer and his apostle Judas with his brother, Sanandana Kumara, better known as Christ Yeshua. Without Sanaka, Christ would have been a mere footnote in history and the crucifixion/resurrection journey would not have happened.

Sanandana Kumara, "the eternal bliss of truth",[237] is the head of the lineage for liberation of the soul into God consciousness through love, joy, devotion, bliss, goodness and truth, and had his most famous incarnation as Christ Yeshua.

He, like all four Kumaras, was conceived and born immaculately in the heavens *and* on earth, conceived by God as a "Son of God". As Yeshua, he was trained in India and Egypt, and is described in the Gospel of Judas as physically appearing like a child (in his true form as a Kumara) before the Apostles, echoing one of his most famous sayings, "Be ye like a child to enter the Kingdom of Heaven."

His journey into God Consciousness, like Osiris before him, was in 14 steps. These 14 are known as the 14 Stations of the Cross, and are celebrated every Easter by billions of Christians worldwide. The 14 Stations tell the story of Jesus' journey from mortality into immortality after he was "betrayed" by his brother Judas (aka Lucifer aka Sanaka Kumara). In the pre-Christian Egyptian story, from where modern Christianity is derived, Judas' role is played by Set, brother of Osiris.

In the Easter ritual commemorating this process of 14 steps to resurrection/enlightenment, 14 images are arranged along a path so one can stop at each station to say prayers or hymns, again a throwback to the earlier Egyptian ritual use of mantras and *heka*, words of power. *These 14 Stations of the Cross are:*

1. Jesus is condemned to death.
2. Jesus carries His cross.
3. Jesus falls for the first time.
4. Jesus meets His mother, Mary.
5. Simon of Cyrene helps Jesus carry the cross.
6. Veronica wipes the face of Jesus.
7. Jesus falls for the second time.
8. Jesus meets the women of Jerusalem.
9. Jesus falls for the third time.
10. Jesus is stripped of his clothes.
11. Jesus is nailed to the cross.
12. Jesus dies on the cross.
13. Jesus is taken down from the cross.
14. Jesus is placed in the tomb, to then, in the 15th step, *become ... Christ in Resurrection, like Osiris before him.*

The 14 Stations of the Cross are known as the *Via Dolorosa* or Way of Sorrows, and are mirrored by the 14 Stations of the Resurrection, the *Via Lucis* or Way of Light. Christianity culminates in, and is incomplete without, the Resurrection. Both of these 14 Stations mirror each other: one spiral of vibration is flowing out, and one spiral is flowing in: the torus spiral in action.

When we cross-reference the 14 Stations of the Cross and the 14 unfoldings of consciousness in the Maheshwara Sutra, we receive a remarkably correlated picture of the transformation/resurrection/enlightenment/creation process.

1. A I Un – The Word, The First Light of The Creator

"In the Beginning was The Word" or the "Light from the Void."

Jesus is condemned to death.

This whole process is a mirror of "as above so below". The beginning of light and creation in the heavens is mirrored and reversed on earth in Yeshua being condemned to death, marking the beginning of his crucifixion and resurrection journey. His mortality is ending and his immortality process is beginning in order to become an immortal human: a God-Man.

The first vibrations of a new creation coincide with the death of the old. This is Shiva's Dance of creation and destruction happening at the same time, mirrored in the heavens and on earth, as above so below, just in reverse reflection.

2. RLRK – One and Maya Creating

"I Am The Light of the World."

Jesus carries His cross.

Yeshua starts his journey into the "underworld", bringing the One Light of awareness *AIUn* into the beginnings of form *RLRK*, Knowing this is the light of the world where "man is made in God's Image". Yeshua Sees this pure light in the matter of his body, and brings light into matter by taking up his cross, the ancient symbol for the union of light and matter.

3. EOn – Primordial Creating

The Nativity

Yeshua falls for the first time.

The "fall" from formless awareness into matter is brought about by the creation of the quantum blueprint for polarities in the Cosmic Egg *EOn*.

4. AI AU – Toroidal Wave-Field Geometry
Baptism
Jesus meets his mother.

Yeshua accesses the fountain of light and life *AIAUch* that blooms forth to birth the worlds. This is his baptism into the vibratory current where the quantum code creates form through its agent of the torus vortex and its space-time geometry field. Mother and birth are synonymous.

5. HaYaWaRAt – Space, Air, Water, Fire
"I Am the Living Bread of Life."
Simon of Cyrene *helps Jesus carry the cross.*

The four elements are the living bread of life, the foundational energy sources and constituents that create matter and physical life. Simon comes to help Yeshua in this unfolding of the four elements that compose his mortal body, so that these elements are reconnected in him to the quantum code.

6. LAN – Earth Element
*The Transfiguration: Veronica wipes
the face of Jesus.*

The light of Yeshua's sweat and blood-stained face becomes imprinted on Veronica's cloth. These are constituents of the earth element, and his vibratory imprint is now recorded, like a photograph, on the "cloth" of the earth element. LAn shows the formless light becoming matter and human, as earth element solidifies all elements into our 3rd dimensional physical body.

7. NaMaNaNaNAM – Elements expand into the subtle senses.

"I Am the Vine."

Jesus falls for the 2nd time.

Yeshua falls for the second time into the vine or elemental thread that connects the elements to the subtle senses of touch/feeling, form, sound, taste and smell. The vine is made of the intertwining threads of the elements and their tanmatra *measures*, which subdivide and expand out the elements, extending their tree of influence and further explicating the matrix of life. These measures or tanmatras connect the elements and subtle senses to the first light of creation and quantum code in an unbroken thread or vine. Paradoxically in the reverse mirror of this story, he begins to lose these supersensory capacities and falls again.

8. JHA BHAN – Speech, Understanding, Inside-Outside

"Do it to the least of you and you do it to Me."

Mt. 25

The women of Jerusalem mourn as Yeshua loses his ability to speak and comprehend what is occurring to him as a human being. He fully experiences the consciousness of being inside and outside at the same time whilst still being situated in God Consciousness, hence his saying here: "Do it to the least of you and you do it to me."

He feels everything, the human frailty and suffering as well as the God within him. There is no division between inside and outside, between him and all life forms. Yeshua affirms that awareness is in every being, and he is part of that, and even though his speech and understanding is dissolving, God is still present, independent of the mind and body.

9. GHA DHA DHASH – Moving, Releasing, Procreating

"I Am the Good Shepherd."

The third fall occurs.

Movement through his legs, the sexual energy and the ability to release, the in-and-out waves of the toroidal spiral, are taken from Yeshua. Thus He "falls" again as his fundamental human procreative sexual power dissolves. He is becoming less and less human, stripped of the basic markers of being a human being.

His three falls are the three intersecting points between light and matter, where the incoming, outgoing and neutral standing waves of the torus meet.

10. JaBaGaDaDash – The five senses

"I Am the Door, the Gate, the Way."

Yeshua is stripped of his clothes.

Yeshua is stripped of the basics of physical 3D life: the five senses, the doorways into the world of matter. He can no longer see, taste, touch, hear or smell. Yet still, eternal formless awareness is present.

The retraction or internalisation of the senses is the basic premise of meditation: shut off your senses in order to access the infinite, as modern yoga shares. Thus, Yeshua states, "I Am the Door, the Gate, the Way," as this is a Way into the formless light.

The senses are the "clothes" we wear in order to navigate in the world, and the clothes the Infinite dons in order to experience and enjoy the world of form.

11. KhPhChThThCaTaTav – 5 Pranas and the Trinity of Mind

"I Am the Way, the Truth, and the Life."

Yeshua is nailed to the cross.

The five pranic flows that sustain the life of the body-mind leave Yeshua's body as he is nailed to the cross. The cross is the ancient symbol for the union of light in matter/the body, meaning that as the five life-forces leave his body, his mind also dissolves.

Yeshua states, "I Am the Way, the Truth, and the Life," as his body-mind starts to die into the quantum code. He is accessing the universal I AM, and he is not stating he personally is the Way, as dogma would have us believe. No, he is stating that The I AM is the Way, the Truth, and the eternal Life, and he is accessing this I AM in a Way he never has before.

12. KA-PAY – Light and Matter
"My God, my God, why have you forsaken me?"
Yeshua's body dies on the cross.
His heartbeat stops, his breath ceases. Yeshua cries out in grief as he dies, a natural, human reaction and his last mortal reaction. Only the light of consciousness remains, the naked, unadorned, formless living light of Purusha. His last breath exits the body that Prakriti made for him to experience the world with. In the divine mirror, light Purusha and matter Prakriti unite and embody in him.

13. SHASASAr – 3 Spirals of Action, Light and Form/ Stability
Yeshua is taken down from the cross.
Yeshua transcends the three fundamental *gunas* or wave functions of light, action and form. These three are primary tools of the architect of consciousness Maya, according to Krishna. With this final act of human consciousness leaving his body, the body becomes 21 grams lighter.

The light of consciousness uses these three gunas to exit the body and the earth's sphere of influence. Here, Yeshua has mastered the three gunas to become like Shiva in the formless living light of awareness. His body enters the final stage of preparation in order to become *eternal light in form* in his impending resurrection.

This 13th movement *sasasar* is a mirror of the first movement of creation *A I Un*: light, action and form. The journey of creation and dissolution happens between these two reflected wave trinities.

14. HAL – The Gateway Beyond
Yeshua's body is placed in the tomb.

Yeshua enters the darkness of the black hole *HAl* alone, just like Osiris. The alchemical birthing period in his cave tomb for this process lasts for three days. He has left the 13 dimensions of vibration and enters the formless awareness in order to embody infinite living light in his body. This light awakens his dormant "junk" DNA.

Yeshua is Reborn and Resurrected as Christ Yeshua, upgraded into Christ or infinite awareness. He has gone through the portal of the black hole *HAl* into the white hole of *AIUn*, and come out of his tomb. He is now embodying the Supreme light in its totality on earth through this process of 14 steps.

Fourteen, 42 and Christ Yeshua have other remarkable parallels, as there are three sets of 14 generations of Christ's ancestral bloodline, equalling 42. As Matthew 1:1-17 shares: "All the generations from Abraham to David are 14 generations, from David until the captivity in Babylon are 14 generations, and from the captivity in Babylon until Christ are 14 generations." So, 14 and 42 directly led to the creation of Yeshua!

Everything relating to the physical body, etheric body and DNA is connected to 42 generations, because development or evolution in time is governed by the number 7. In the Essenic wisdom, a spiritual group which Christ Yeshua was a part of in ancient Palestine, it was said that one has to pass through 6x7 stages = 42 stages.

This 6x7 is connected with the Earth, and what lies beyond the 6x7 leads beyond earthly, mortal and genetic conditioning.

The next seven steps complete the process of[238] evolution of the physical, etheric and genetic bodies, achieved at 7x7 generations. *In these final seven generations complete transformation takes place and* something new is born: enlightenment. This is what Christ's Crucifixion and Resurrection journey completed for him.

A further validation of this 14/42 process is revealed through the Hebrew alphabet of creation,[239] which forms from a vortex with 14 features, as proved by Stan Tenen. In Kabbalah, God creates the Universe through the 42 Letter Name of God found in the *Ana Bekoach*, "God's first gift to Man and our most important one," as Kabbalistic scholar Jeffrey states. As Rav Brandwein, the Admor of Strettin further shares, "it is through the understanding of the 42 Letter Name that the *Geulah* [final redemption] will occur." This 42 Letter Name is comprised of **14** Triplets, and Pi breaks down into the first 10 Triplets.[240]

Thus we now find the 14 Stations of the Cross and 42 genetic steps of Christ that led to his awakening, the 14 Classes of Sound and 42 syllables in the Maheshwara Sutra depicting creation and awakening, the 14 body parts of Osiris and the Egyptian 42 steps to enlightenment, Egypt being separated into 42 kingdoms as a mirror of the heavens, the 14 features that form the vortex of the Hebrew Alphabet *and* the 42 Letter Name of God in Kabbalistic teachings through which "redemption" occurs.[241]

All of this is no coincidence!

This ancient pathway to enlightenment in the Vedic, Egyptian and Judeo-Christian traditions is further corroborated in TianTien Buddhism, which has 42 Stages to Enlightenment. The first 40 stages of its Bodhisattva Path are "The 10 abodes of inspiration", "The 10 practices of virtue", "The 10 transfers of merit", and "The 10 groundings of enlightenment". Lastly, there are the two levels of "Complete, Universal Enlightenment and manifestation of the Buddha" and the "Original, Eternal Buddha", or *HAI*.

Curiously enough, the number 42 has also made it into modern culture with the best-selling book *The Hitchhiker's Guide to the Galaxy* by Douglas Adams. The number 42 in the book is the *"Answer to the Ultimate Question of Life, the Universe, and Everything,"* and is calculated by an enormous supercomputer named "Deep Thought" over a period of 7.5 million years.

Resurrection

"I Am the Resurrection and the Life."

In the beginning of the Resurrection process, One Sees, *like parasakshi*, the created universe of forms, shapes, vibrations and appearances in the supreme detachment of unconditional love. Clearly discerning Maya's workings, one feels the love in all beings as part of one's own Self, and through the black hole of *HAl*, you return back into creation through the white hole of *AlUn* now established in the Infinite.

You are resurrected, twice born. The first birth was from your mother's womb and the conditioning of your parents, ancestors and your own karmic conditionings; this second birth or *dwija* is when you are born from the infinite without any conditioning, born through the supreme detachment of unconditional love.

You now appear like a human being, and have the same structures of every human being: yet you are now *beyond, Para*, the human condition as a conscious Particle of the Infinite Field. You Realize that the whole of the universe is now contained within your Temple: and you are the same living light in all beings.

Creation and experience has a beginning and an end, but the fundamental awareness that all creation and experience arises from never ends as it has no beginning. Energy that is not confined has no limit, no form, no name and no vibration as we define vibration. It is infinite and does not move because there is nothing beyond infinity. It is not defined or limited by time and space, and has no boundaries, no "here" to differentiate from "there".

The dissolution and resurrection of a person's consciousness into *HAL* means you no longer independently think or do anything that is not of *HAL*, the Supreme Witness. Your consciousness is now fully participating in the all-knowing infinite continuum of energy found in the ever-present Now.

The I Am *Aham* is the Lord of creation *Isvara*. This is YOU. Aham is the formless in form, coming from *Maha-Ishwara* (Maheshwara), the Supreme Aham I AM. Similarly, Yeshua became The Christ or Anointed One after his initiation when he became Christ Yeshua, Maha *and* Ishwara as One, echoed in his saying: *"I and my Father are One."*

In his Resurrection Christ shared: *"Ye are all like gods and all of you are children of the Most-High"* in John 10:34 and Psalm 82:6. Christ took this "hero's journey" of 14 steps of initiation into the One. His entombment as Yeshua in the 14th step ended when he was seen and aided by his wife and shakti Mary Magdalene into the fifteenth step of resurrection.

Similarly, his predecessor Osiris in his 14 movements was resurrected by his wife and shakti Isis and given the Lingam, the regenerative power and symbol of Shiva, in his fifteenth step. Both Osiris' and Christ's journeys hearken back to their founding myth in the *Purusha Suktam* of the *Rk Veda*, where the Cosmic Soul *Purusha* sacrifices his body into many pieces in order to birth creation into being.

Christ has always been associated with Venus, the home of the Kumaras. Venus is known for its dual aspects of morning and evening star, which have been seen throughout time as brothers in Sanandana and Sanaka Kumara, Christ and Lucifer, Yeshua and Judas: all twin aspects of the same One.

In Vedic lore, Venus or *Sri Shukracharya* (shukra means semen) is the guru of the *asuras* or demons. In the *Devi Mahatmya* and other Vedic texts, Venus possesses the *sanjeevani vidya* – the power to resurrect the dead. Venus used the sanjeevani vidya to resurrect the members of the demon armies whenever they died, perpetually giving them an advantage over the armies of the gods or *devas* in their long-standing war.

The sanjeevani vidya was a process greatly treasured and guarded by Venus, which the *devas* tried to acquire many times through fair means and foul, praise and trickery.

Mary Magdalene was a High Priestess of Isis trained in Egypt and heralded in the Western esoteric tradition as being from Venus; Isis was the sister of Hathor, Goddess of Venus. Both Magdalene and Isis may have possessed the *sanjeevani vidya* and used it to help their masculine partners, their Shivas, into their resurrection into cosmic consciousness.

And who, may you ask, originally gifted the sanjeevani vidya to Venus? *Shiva.*

In using the sanjeevani vidya, both the "light" and "dark" were used to bring one beyond duality: *advaita.* Yeshua and Osiris went through their enlightenment processes *as partners* to Magdalene and Isis, their Shaktis, just as the universe could only be created through Shiva and Shakti, as seen in the four-mated masculine-feminine pairs revealed in the dance of the Maheshwara Sutra.

Shiva Jyotir Lingam

What Shakti gives to Osiris and Christ in order to resurrect them is the Lingam and/or the sanjeevani vidya. No one knows exactly what the sanjeevani vidya is: it could be a mantra, a spell, a potion, something connected with the Lingam, or all of the above.

If we assume Venus, as the god of beauty and sensuality, is connected to the powers of the lingam and lovemaking, and that the sanjeevani vidya was given to him by Shiva, it is fair to assume the resurrection energy could be the Shiva Lingam distilled into an alchemical form designed to resurrect life after the Hero's Journey of these 14 Steps.

The *satya stambha* or Shiva Jyotir Lingam is an eternal and infinite pillar that stretches from the top to the bottom of the tree of life, and beyond it through *HAL*. This lingam is what Isis gave/initiated Osiris into as the final part of his resurrection, and is mentioned prominently in their story. It may be the same power which Mary gave/initiated Jesus into after his physical death as he resurrected into Christ Consciousness.

In the story of the Shiva Lingam:

It is said that at the very beginning of creation Brahma and Vishnu met each other in the vastness of primordial nothingness. Each proclaimed their greatness, asserting they were, in fact, the One and Only Creator of everything.

As their debating went on, suddenly, in the space before the two gods, there appeared an immense flaming pillar of fire and light, a burning, illuminating presence they did not know and had not met before.

Astonished by this, Visnu changed his form into a boar to investigate where this pillar came from. For a thousand years, he tunnelled down to try and find the bottom of the pillar, to find its source. Diving deeper and deeper he travelled, only to return back again having not found the beginning to this flaming pillar.

Similarly, Brahma, amazed in a fit of piqued pride, changed his form to that of a bird. Ascending higher and higher on his wings, he travelled for a thousand years to find the top of this pillar, only to return back again having found no end to this flaming pillar.

Brahma and Visnu met again at the flaming Shiva Lingam, crestfallen and bewildered. Suddenly, Shiva appeared before them, reverberating and rippling the fabric of space, blinking them in and out of existence like a mirage.

Vishnu and Brahma did not need to say anything, for with one glance they both knew. Heads down, they folded their hands and bowed, upon which Shiva acknowledged them both with a gracious, knowing, radiating smile of light soaked in bliss.

Shiva in the form of the Jyotir Lingam is a terrifyingly powerful, magnificently unstoppable force, coming out of the immeasurable depths of nothing in living light. The Shiva Lingam penetrates everything, inhabits everything, becomes everything, and fills up everything with Itself. From it, everything arises.

Shiva's Hologram

Of course, Shiva Lingam wants to be with and meet its yoni, its Shakti, as we see in modern day Shiva temples where the Shiva Lingam is embedded into the yoni as One Flow. This mirrors the universal creative process, as every reflection that unfolds creation is in complementary pairs.

All creation unfolds through waves, but in quantum holographic physics, waves of light do not travel. *They mirror each other from wave-field to wave-field of space.* The planes of zero curvature which bound all wave-fields reflect light from one field to another.[242] Each reflects the other.

This is vividly echoed in the vision of the holographic *Indra's Net* of ancient India, where reality is described as: *"an imperial net made of jewels. Because the jewels are clear, they reflect each other's images, appearing in each other's reflections upon reflections, ad infinitum, all appearing at once in one jewel, and in each one it is so."*

The hologram of creation manifests through these "jewel" nodes of seed syllables mirroring each other (as seen throughout the Sutra). All creation is formed through sounds/waves/Shiva-Shakti reflecting each other in the mirror of the One Consciousness.

In a hologram, the whole pattern is in every part. If you were to take out any part of the hologram, it would be whole and complete in itself *and* contain every other part of the hologram. If you were to take out any part of the hologram, you would see the whole pattern repeating itself again and again. To change one part of the hologram means altering the rest of the hologram.

Within each thing lies every other thing. Within each of your cells lies your entire body, and your parents, children, grandparents and grandchildren even if they are not yet born: similarly, within the seed of a tree lies the tree and all generations of that tree.

Shiva's Hologram is a reverberating vibrational matrix, with all 42 sounds in 14 pulses simultaneously and perpetually resounding all the time in every moment in creation *and in you*. Can you imagine this Sound?

This surround sound *Spanda*, repeating over and over again every second, is the sound of 14 waves of virtually endless harmonic permutations that compose the total range of perceptions, forms and transformations possible in creation.

This Sounding is an intensity of inclusiveness, the sheer potency of all creation resounding all its sounds and frequencies at the same time. This experience is literally like being born and re-born in every single second, and is an incredibly powerful experience.

Only the people who have gone beyond the world can change the world.
– Sri Nisargadatta Maharaj

So: If you had the power to create a holographic universe, would you? Or wouldn't you? Creating a black hole is the precursor to creating a galaxy, and a mega black hole could birth a universe through a "big bang" or more accurately a "silent clear light", seeing everything into being.

Black holes are relatively easy to make. The first step is understanding that for any object, there is a critical radius at which its mass will form a black hole. For the Sun to become a black hole, its radius would have to be squeezed into a size about two miles wide.

Today, we have the technology to create black holes through particle accelerators, which squeeze masses into tiny volumes.

The best thing is that since gravity has negative energy, it takes very little energy to make a black hole.

The second step involves nudging the properties of the black hole, setting precise parameters and designing it in detail according to what the purpose of the creation would be, as you saw fit. Like with designer babies, instead of tinkering with DNA to get a "perfect" child, a Designer would tinker with the laws of quantum physics to create a universe according to whatever they wished to recognise, experience or accomplish.

Life as seen by these Designers or Gods is much different from our conception of it. Such a Designer is Siva, with his mantra *Sivohum*: I Am Siva,[243] and *Ahamatmo Paro HAL*.

Embodiment

Embodying on earth with a human form after Resurrection comes with a simple understanding. Awareness is That which all experience arises within, which all things arise within. In this ever-present awareness, all waves of consciousness appear and dissolve. Awareness is untouched, unmoved by these waves.

The uninterrupted continuum of these waves of consciousness incessantly streaming forth in every nano second throughout everything in all creation is Shakti. The body and nervous system are designed to be transparent to these waves of consciousness. These waves create time: time forms bodies and death.

Awareness is what all experience and waves arises within. Walking, speaking, acting from awareness means you are acting so totally that "you" are absent, you are so engaged in all that you do that you are not here. 100% = 0%.

The body-mind arises within awareness. Awareness is not in the body-mind. There is no "coming into the body". To have a transparent body-mind and nervous system to the waves of consciousness, which flow throughout everything, is part of the design of Awareness. There is no resistance to this infinite continuum of waves incessantly streaming through everything.[244] There is no accepting or embrace in This. It just Is.

It is said that when one is embodied, their auric body or *kosha,* that is the etheric blueprint of the physical body, looks like a 25-year-old. If you were to see an awakened person as they truly are in their auric body, not their physical body which may be old, grey, wrinkled, they would look around 25. You can see this if you look with the eyes of the soul. If a person is fully embodying Awareness, their spirit body looks like an innocent child.

When one is not embodied, the spirit body looks old, broken, has missing limbs, a lopsided face, distorted body, even if the physical body is young and fit. The six-pack-stomached man on the beach, the lithe toned woman, may actually look 60 with a pot belly and a limp! Conversely, the 60-year-old embodied one who walks slowly, with grey hair, has a spirit-body six-pack stomach, toned physique, long luxuriant hair, and radiant light.

Embodying This here and now entails physicality, an aliveness, an engagement, a goodness, a vitality, a being with, enjoying all facets of experience here and now because you are total in it *and* absent from it in the supreme detachment of unconditional love.

You enjoy it all totally, and in this enjoyment, you are totally absent: then you are embodied. This is the Tantra of the Maheshwara Sutra.

The Flower of Life

The Flower of Life was an ancient school of wisdom established many thousands of years ago in ancient Egypt,[245] and was intimately associated with Osiris. A large Flower of Life geometry is carved into the wall at the Osireion, his main temple in Abydos.

The Flower of Life then spread across the ancient world, with its geometric pattern having been carved, painted, drawn and burned into stone in over 60 countries, including all across Europe, China and India.

In India, the Flower of Life Geometry has been most notably found in Kashmir in the mountains outside Srinagar, where Moses' Tomb is alleged to be. Srinagar itself is said to be where the tomb of Christ lies, in a non-descript building in the heart of the city. Many accounts from ancient India, including one from the court of the King of Kashmir, describe Christ Yeshua as living in Kashmir.

The quantum matrix of creation described in the Maheshwara Sutra is remarkably similar to the Flower of Life creation geometry sequence, both of which show the unfolding of creation from the beginning of the universe.

Both are almost identical in every step of creation they map out, with the Sutra providing the sounds and precise actions, functions and meanings for the geometric progressions of the Flower of Life. Together, the Maheshwara Sutra and Flower of Life map out the universal language of creation in sound, geometry and number.

For example, the first 6 movements of Creation in the Maheshwara Sutra are *AIUn, RLRK, AON, AIAUch, HaYaWaRAt*, all of which coagulate in the sixth movement of Earth *LAN*.[246]

In the Flower of Life, these 6 Circles are seen as the 6 Days of Creation, as mentioned in the Biblical Book of Genesis. The Sixth Day is when the formless is brought into form and everything is created, which in the Sutra is the creation of earth element: when the formless has been brought into form.

The geometry of the Sixth Day is called the *Seed of Life*. This Seed of Life is the Blueprint for all life to be brought into form from the formless. It is the same 6 steps in both traditions, as we can see from the diagram below.

The Seed of Life Unfolding

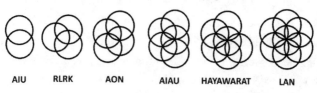

AIU   RLRK   AON   AIAU   HAYAWARAT   LAN

The first 6 classes of sound in the Maheshwara Sutra are the first 6 unfoldings of the Flower of Life that form the Seed of Life, which then continue expanding to form the fully explicated and completed Flower of Life as we saw in the chapter on *GhaDhaDhash*.

The Seed of Life consists of 7 overlapping circles with the same diameter. Six are regularly spaced within the 7th, producing 18 petals or vibratory lenses: 6 smaller ones inside and 12 larger ones outside. These are 18 sounds, as the Sutra shares.

In the Sutra, *A I Un* creates the 1 and the 2 interconnecting circles of the Vesica Piscis; then the subsequent 5 classes of sound unfold the 18 syllables of the first 6 classes of sound:

A I Un has 3
R-LR-K has 3
EOn has 3
AI AU ch has 3
HA YA WA RAt has 4
LAn has 2.

The first 6 classes of sound in the Maheshwara Sutra are the first 6 unfoldings of the Flower of Life that form the Seed of Life.

These 6 unfold 6 lines or *6 directions* emanating outwards. These 6 directions have been used by cultures around the world for thousands of years to orient themselves to the natural order of the earth and human being in harmony.

As St Clement of Rome shares: *"Looking forth (on these lines) as a number equal in every direction, God completed the world in 6 equal divisions of time. For at Him the six boundless lines do terminate, and from Him they take their boundless extension."*

Similarly, this process is recounted in the Hymn of Creation, *Rk Veda*: *"Out went their lines, crosswise; was there above, was there below? From below pure potency for Self Birth. From above the force to move, rise, and fill."*

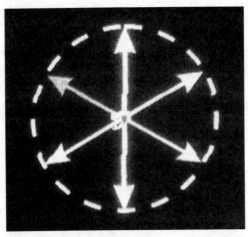

Spirit projecting consciousness into the six directions. From **The Ancient Secret of the Flower of Life** *by Melchizedek.*

The six lines are infinite emanations of the One awareness, the actions of the quantum code of the Maheshwara Sutra and Flower of Life. They are awareness in movement, the consciousness of awareness. *A* is where all six lines arise from. There is no difference between the six lines and awareness/void, just as the rays of the sun are not different from the sun itself.

These six directions are described in quantum physics as *Null Lines*, the paths taken by light rays and mass-less particles. A geometry based on null lines is the Grail of quantum physics, for in a universe having such geometry, mass does not exist.

All null lines always have 0 length: scale and distance are also 0. In other words, all distances, times, ideas of large or small, micro and macro, no longer exist, and therefore no time would elapse from travelling from one point to another.

In terms of light, not even one second passes from the time a ray of this light leaves, say Orion, to Earth, for along such a null line the distance to the stars is 0. When you look along a null line, nothing separates you from what you "see" in the Universe around you. In fact, you are everything around you – there is no separation.

Meditation

To visualise this, if we look at the five emanating lines or patterns in the Seed of Life, we only get one hemisphere of the world around us. To complete the full picture, turn around to look BEHIND you to find the last line-pattern. When this is seen, you have a 360-degree picture in all directions: a complete sphere.

The six lines are unlimited. Visualise this. Imagine the focus or central point of the lines being your heart. One line goes up to the sky, one down to the core of the earth, one to the right, one to the left, one in front, one going behind you.

The Hexagon

The Seed of Life is 6, the number of sides of a hexagon. The 2nd hexagon within the 3 hexagons that make up the fully explicated and complete Flower of Life has 42 Petals within it, as you can see below.[247] This of course correlates to the 42 syllables of the Maheshwara Sutra.

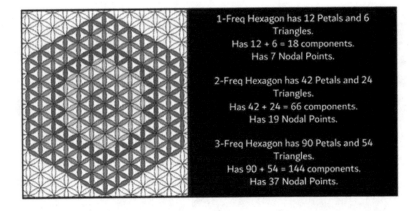

1-Freq Hexagon has 12 Petals and 6 Triangles.
Has 12 + 6 = 18 components.
Has 7 Nodal Points.

2-Freq Hexagon has 42 Petals and 24 Triangles.
Has 42 + 24 = 66 components.
Has 19 Nodal Points.

3-Freq Hexagon has 90 Petals and 54 Triangles.
Has 90 + 54 = 144 components.
Has 37 Nodal Points.

The geometry of 6 is a hexagon, which holds vibration in its most efficient transmitting and receiving arrangement, or packing order. All space is utilized in full; no gaps arise. The hexagon achieves the maximum storage power in the minimum of area and space. This is because the radius of the circle can walk around its circumference 6 times.

It is a highly balanced system, and perhaps this is why Carbon, the most important element for compounds necessary for earth life, has an atomic number of 6, with 6 nucleons, 6 protons and 6 electrons.

The hexagon is the perfect holder and carrier of vibration, as seen throughout Nature.[248] A hexagon forms seed pods, shown in the way that bees organize their communities in hexagonal cells, and the way that water crystals and snowflake geometries form, showing how water crystallises from liquid to solid, choosing 1/6th of a circle or 60 degrees to do so.

The hexagrams of the Chinese I Ching and the six-sided crystal of the Earth's core echo this resonant geometry, and the hexagon is used worldwide by architects of sound chambers, cathedrals, temples and vibratory architectures in India, Egypt, Europe and in Native America.

In Native American lands, dome-shaped Sound Chambers are used to attune to the universal forces of the 6 directions through the 6 Vowel Sounds, as popularised by Native American Elder and author Joseph Rael.

The architecture for these sound chambers is based on the first sound of creation A and the geometry of the six directions. As dowser Richard Crutchfield PhD describes, these Sound Chambers receive shafts of energy from each of the four directions, *and* vertically from above and below, making 6 directions in all.

Curiously, recent advances in acoustic technology have resulted in the creation of hexagonal speakers operating in 6.1 Surround Sound by Funktion-One Audio. This sound system creates an all enveloping field of sound, so that wherever you go within the field the sound is the same and is tactile, i.e. you can feel it with your skin and electromagnetic field![249]

The Forming of the Embryo

The 42 sounds and geometries of creation are 42 steps unfolding the science of creation. The 42 show the universal creative pattern *and* the development of the embryo as the mirror of this creative pattern. In the great saying: "as above, so below", what happens "above in the heavens or macro universe is mirrored below", in the micro universe of our bodies.

We contain the universal patterns within us. By looking out to the universe we are actually looking at ourselves and our own patterns. From the nothingness of the void the universe sprang forth; so there is a void in the human womb until a foetus is seeded there.

The physical body is formed by the organizing forces of the embryo, which manifest at conception through the quantum field. These organizing forces diffuse into the fluids of the embryo as an embodied ordering force. These embryological forces manifest through toroidal wave-fields *within and around* the body which form and direct the fluid tides within *and* around the body, manifesting within the body as forces within the fluids and as tidal motions which permeate cells and tissues.

All these toroidal wave-fields follow the quantum blueprint established for their creation in *AIUn*, which then bloom into the manifested torus wave-field in *AIAUch*. However, as we grow into adults these wave-fields become distorted due to our conditioning.

The conditioning we experience in our formative moments of early life generate form *and* protective responses, such as tensile patterns in our fluid and tissue fields. This then becomes part of the organizing forces of the trinity of the mind and its defensive processes.

To re-establish right relationship of the mind, pranas, cells and tissues to the original quantum blueprint of the toroidal embryological matrix is important! Understanding how this works is one step towards this reordering and realigning to our quantum blueprint, as these embryological forces and organization work in the adult system.

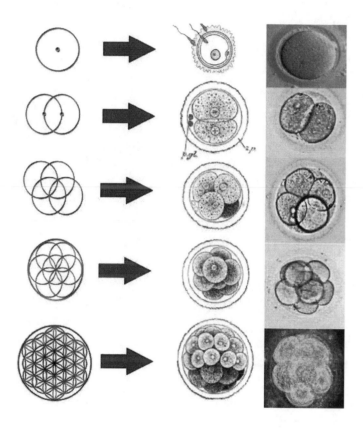

The Sequence of Creation in the Foetus

With all 42 steps of creation completed, the embryo is fully incarnated with all of its hardware embodied into a physical form.

A – The First Light: Day 1 of Foetus Forming

Around the ovum lies the liquid membrane of the zona pellucida, and inside this lies the sphere of the female pronucleus, which contains 22+1 chromosomes. 12 male sperms surround the Ovum in a masculine cuboctahedron grid, all 12 sperms touching the centre sphere of A.

I – Activating: Day 1

One sperm penetrates and fertilizes (activating through the I energy of Shakti) the ovum, becoming a perfect sphere itself, the same size as the female pronucleus and containing the other half of the 22+1 chromosomes. These two spheres pass through each other in the geometric form of the Vesica Piscis.

Un – The First Cell: Day 1

As the two pronuclei make the Vesica Piscis, the 2 become 1. The zygote or first cell is created. This is the third part of the holy trinity: the individual you or Isvara. The 1 becomes the 1, to become 2, then marries itself within indivisible quantum unity to form the first "holy" trinity that begets you.

A as the 1 Just Is; then the 1 Is the 1 in I, the 1 is the 2 in A-I, and the 1 is the 3 in **Un**. These are 4 movements. A I Un is an indivisible quantum power of 3 vowels + 1 consonant n, n being the quantum matter of existence. These 3+1 movements are part of the quantum field and are indivisible in Day 1 of the creation of the foetus.

R LR K – Consciousness Forms: Day 2

Two cells multiply into 4 and a star tetrahedron forms. The 4-celled pyramidal tetrahedron is the 4 centres of 4 spheres.

There are 27 smaller tetrahedrons nesting within it. The tetrahedron can turn inside out, as it is a fractal. Inside is the same as outside. This is the mirror of reflection RLRK, where consciousness becomes aware of its source in the same field of One.

EOn – Immortal Egg: Day 3

Four cells become the 8 original cells at the base of our spine in 2 interlocked tetrahedrons, one male pointing up, and one female pointing down. The 2 interpenetrating tetrahedra are the tantric union of masculine-feminine forces, where polarities unify in balance. Two poles form, with the centre point the perineum of the foetus.

This geometry forms a Cube of our 8 Original Cells, the code and template for our Incarnation. These 8 Cells never change from the moment of our conception to the day we die. Every other cell gets replaced every 5-7 years, but not these original 8.

These 8 Cells store the memory of your Purpose, why you have come here, your evolutionary lessons and karmas, and your entire history. Realigning to these 8 Original Cells can also serve to get you back on your pathway if you have been distracted, helping you claim your direction, power, purpose and guiding light within.

From these 8 Cells the foetus begins to grow radially in a more feminine 360-degree expanding pattern, again mirroring the process of the egg of EOn expanding and developing into *AIAUch*.

AI AUch – Creation Blooms: Day 4

Eight cells become 16 in two cubes. Metatron's Cube Geometry is formed, a cube within a cube, where inside and outside are the same. After this, the previous order of geometric symmetry dissolves as a new order of creating begins in *AI* through the toroidal spiral.

This marks the movement from the quantum field into physical manifestation. The toroidal wave *AI* moves through the foetus in order to form the body within the organizing space-time geometric field of *AU*. Both masculine-feminine polarities are in union, and inside and outside are the same.

HA YA WA RAt – The 4 Elements: Day 5

As the foetus expands into 32 cells it becomes a sphere again, further unfolding into 128 cells then 512 cells, 8 cubed. A hollow fluid-filled space forms like an apple through the motion of the torus spiral, which turns inside out and is fully connected to itself: in and out are one flow. This is the *morula* stage of embryonic development. The hollow space becomes the heart, then the tongue emerges from the heart.

The driving energies of the torus at this stage are composed of the four elements Space, Air, Water, Fire, through its spiralling flow.

LAN – The Earth Element: Day 6

The blastocyst sheds the membrane surrounding it and embeds into the womb (the earthing element) in order to further evolve. *Hatching occurs*, as the earth element grounds the other elements into form. The Seed of Life is formed as the core blueprint geometry for the foetus to settle into itself and further develop from.

NA MA NA NA NAM – The Subtle Senses: Day 7

Creation here is an emanation of the elemental powers and their threads of vibration that expand outwards from the Seed of Life to form the subtle senses of sound, form, feeling, smell and touch, all of which are created as ways for the Self to recognise and express itself.

JHA BHAN – Speech and Understanding: Day 8

The toroidal wave forms the mouth and capacity for speech, allied with the ability to neurologically string together speech and thoughts. The forming of the hands themselves happens. The neurological function that discerns the difference between inside and outside occurs.

GHADHADHASH – Moving, Releasing, Procreating: Day 9

The anus and sexual organs are formed by the toroidal wave. In conjunction with the previous forming of the mouth in JHABHAN, we now have 2 holes in the body for the in-and-out waves of the torus. Growth goes into itself, and then goes out, contracting and expanding.

The forming of the anus and reproductive organs occurs in one creative motion. With this comes the sexual energy and the capacity for us to reproduce, generate and become creators ourselves. A new blueprint pattern of mortality and physical death is introduced into the evolutionary process, whilst still being connected to the infinite quantum code of the One.

JA BA GA DA DASH – The Five Senses: Day 10

The 5 sense organs form in the foetus, establishing the new code of physicality, mortality and death into the evolutionary process. One now has the hardware to experience temporary life and a temporary world.

KHPHCHTHTHCA TA TAU – The Five Pranas and Trinity of Mind: Day 11

The five flows of life-force fuel the trinity of the mind. The toroidal spiral creates them both at the same time in the same field of consciousness. These five flows animate the body-mind so the mind can now perceive and analyse data through the senses.

One has the hardware to fully identify and function as an individual ego *ahamkara*, in order to live one's individual karma. The awakened or *Buddhic* mind is also formed here, meaning the capacity for *free will is formed here*, as is the capacity to guide oneself in life as a sovereign being connected to the quantum code through the Buddhic mind. OR, one could guide oneself in life through the ego ahamkara. The choice is yours, and this point in the creation of the foetus gives you the choice.

KA PAY – Matter and Light: Day 12

The formless living light or "soul" now settles into the body-mind, taking its place as its ruling power. The body-mind is now enabled to fully align to this living light. The raw energy-information of the body is in harmony with this, as individual soul sentience develops.

SHA SA SAR – The Three Spirals: Day 13

All foetal functions coagulate fully, with the embodying of sentience and its capacity to live and experience the world activated by the three *gunas*. These three wave aspects weave their way through every part of the foetus, grounding and stabilizing functions and performing these actions whilst guided by the illuminating power of sentience.

The three gunas run through the 3 channels of the spine in ida, pingala and the central channel sushumna, completing the activation of the nervous system. Every function in the foetus is now endowed with sentience, so one is prepared to experience the world as a sentient being.

HAL – Completion: Day 14

The living light of supreme awareness leaves the form it has just created, as its role is complete in establishing Its awareness in form. This also creates a pathway for us to be able to do the same: to access this supreme awareness when we are ready to awaken into This.

In India, it is known that it takes 14 days for the sentient soul to leave the body upon physical death as it moves through multiple dimensions of consciousness in the *Bardo* journey. The 15th day is celebrated as the release of the earthbound soul, as it has now gone into its next phase of development, be it liberation from the cycles of reincarnation, or reincarnation back on earth again...

Vitruvian Man and 14

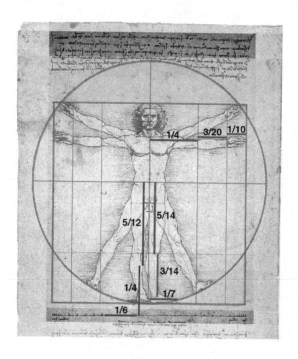

Da Vinci's famous drawing of the Vitruvian Man demonstrates the harmonious proportions and dimensions of the human form, bringing together artistic, scientific, mathematical and sacred geometric knowledge.[250]

Da Vinci was commissioned in 1509 to illustrate the book *De divina proportione*, a book on mathematics and sacred geometry based on the Phi Golden Mean Ratio. This is the ideal heavenly ratio which describes the symmetry of harmonious perfection, which our bodies are close to, but not exactly harmonious with.

In the text penned in mirror writing below the drawing, he says that the ideal human body dimension, in perfect proportion, *is subdivided into 14 parts*, based on whole number proportions. These are:

The length of a man's outspread arms is equal to his height.

1. From the roots of the hair to the bottom of the chin is a tenth of a man's height.
2. From the bottom of the chin to the top of his head is one-eighth of his height.
3. From the top of the breast to the top of his head is one-sixth of a man.
4. From the top of the breast to the roots of the hair is the seventh part of a man.
5. From the nipples to the top of the head is the fourth part of a man.
6. The greatest width of the shoulders contains in itself the fourth part of a man.
7. From the elbow to the tip of the hand is the fifth part of a man.
8. From the elbow to the angle of the armpit is the eighth part of a man.
9. The whole hand is the tenth part of a man.
10. The beginning of the genitals marks the middle of the man.
11. The foot is the seventh part of a man.
12. From the sole of the foot to below the knee is the next part of a man.
13. From below the knee to the beginning of the genitals is the next part of a man.
14. The distance from the bottom of the chin to the nose and from the roots of the hair to the eyebrows is in each case the same, and like the ear, a third of a face.

In their mathematical explorations, Vitruvius and Leonardo were looking for the perfect proportions of man AND of all creation. Leonardo found them, and expanded on the work of

Vitruvius, only using half of his 22 measurements because he had discovered the Flower of Life geometry which gave the final calculations. His drawings and explorations on the Flower of Life are found in the book *The Unknown Leonardo*, which comes from his notebooks.

In another of his notebooks from 1492 Leonardo wrote: "By the ancients, man has been called the world in miniature; and certainly this name is well bestowed, because, inasmuch as man is composed of earth, water, air and fire, his body resembles that of the earth." In other words, man is a microcosm of the universe and "man is made in god's image".

These 14 subdivisions of the human body in perfect proportion show that the human body is a fractal of the entire universe, a truth repeatedly affirmed throughout the Maheshwara Sutra and indeed all Vedic literature, where the Purusha, the perfect man and universal creator, holds the entire universe within his body. Each of us has the universe within us, as above so below, and this can be proved through sacred geometry and mathematics.

Drunvalo Melchizedek expands on the 14 further in his book *The Ancient Secret of the Flower of Life* where he geometrically and mathematically proves that the map of Christ/God Consciousness is found in a geometric form composed of 14x18, 14 again being the key number.

He also shares his findings on the 13-chakra system, using the knowledge of the 14 Vitruvian Man subdivisions to locate these chakras. For example, the nose and chin are featured in Leonardo's Vitruvian Man as key proportions of the human man, which Melchizedek then used to locate the real positions of the chakras. He corroborated these geometric findings by using a Microwave Emissions Scanner to detect the actual energetic emissions of these chakras in the body.

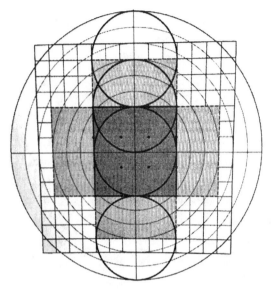

The Christ consciousness: The 14 by 18 square-circle relationship of the third level of consciousness. From **The Ancient Secret of the Flower of Life.**

To add another twist to this, Oxford scholar and author Ashish Jaiswal states that an ancient Vedic text *Vishnudharmottaram Puranam* contains the first ever extensive measurements of the perfectly proportionate human body in the *Hans Man* which describes these harmonious measurements for both men and women, whereas Leonardo only writes about the perfected male form. The Vishnudharmottaram Puranam describes five types of perfect men, and five types of perfect women, all of varying heights and proportions.

In the Vishnudharmottaram Puranam text, Rishi Markandeya explains to King Vajra how to create perfect representations of gods and goddesses using mathematics for the proportions of their bodies. These mathematical measurements are more sophisticated than Vitruvius, giving measurements for smaller body parts (for example, the length of the front tooth of a proportionate human is half of their finger wide).

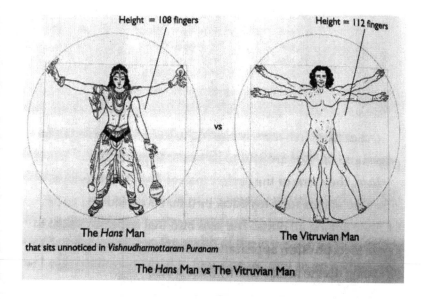

The *Hans* Man vs The Vitruvian Man

Conclusion

The Maheshwara Sutra is the original and most comprehensive teaching on the science of consciousness and creation through sound vibration that has ever been recorded. It has been obscured by time and known to only a few in the Saivite lineage in India *until now*.

This wisdom has been used as an experiential pathway to enlightenment by some of the greatest figures in human history, from Shiva to Patanjali, Christ to Osiris, the original Rishis and Avatars of Vedic India, to Kabbalists and the spiritual schools of the ancient world. All of these initiatic wisdom and spiritual practices reflect the innate foundation of creation, found in 14 and 42.

The Maheshwara Sutra describes how the universe is created from the quantum field to come into manifestation through the dance of Shiva and Shakti, just as how the foetus is conceived to then develop and be born from its parents.

After my first reading of the Maheshwara Sutra, I realized it was not just a Sutra about sound: it was a "Theory of Everything" long sought after by physicists and spiritual scientists alike.

To accurately convey and unfold its many layers of nested meaning required a comprehensive cross-disciplinary approach across all fields of human endeavour, including the sciences of consciousness from various traditions, quantum physics, Sanskrit, Vedic spirituality, linguistics, art, philosophy, vibrational science, sound healing, mantra, sacred geometry and the latest discoveries in cutting edge science and technology.

It also required an experiential and somatic approach to directly experience the heights and depths of consciousness and awareness, achieved through many different spiritual practices, meditations and Tantras.

The Maheshwara Sutra is Shiva's Blueprint for Creation, Shiva's "Theory of Everything", a comprehensive science of consciousness that has cosmic and human relevance. It is practical in its applications in many fields of human discovery and endeavour, from sound healing, mindfulness, yoga, chanting, mantra, to biology, natal therapy, linguistics, the true working of the quantum field, cosmology, the understanding of self, Self and other in relationships, and much more.

The practical use of its sounds with sacred geometry, meditational techniques and spiritual insights about the nature of reality can lead to expanded states of consciousness, and allied with understanding and grace can support one into enlightenment.

The Maheshwara Sutra and its decoded wisdom can also be used to power Artificial Intelligence and new advances in quantum computing hardware and software, as well as generating code for harmoniously generative architectures and engineering advances that can create new types of machinery. All of this and more can lead to humanity's ability to create on a grander scale than has been seen on earth in the last 13,000 years.

The Maheshwara Sutra is non-dual: there is no separation from Creator and Created. Each step in creation, each aspect of creation, is not separate from the One. This is one key to enlightenment held in the Sutra. To live happily in non-dual Knowing whilst still enjoying all that creation offers.

There are many ways in which the 14 classes of sound and 42 syllables can be put together. For example, the 14 classes of sound act as *mirrors of sound*, reflecting each other in 1-14, 2-13, 3-12, 4-11 and so on. These mirrors of sound give a fuller explanation of creation's processes, as within each part of Shiva's Hologram lies every other part. It is not linear, although it is at the same time.

For example, *AIUn*, the first sounds of the white hole birthing creation, are reflected in *HAI*, the last sound of the Sutra and a black hole: 1-14. *RLRK*, the Witness in the 2nd class of sound, births creation into manifestation through Maya, whose main tool is the three gunas *SaSaSar*, which create the basis for every wave in creation and is the 13th class of sound: 2-13.

With the creation of the Cosmic Egg *EOn* comes the revealing of sentient light and quantum matter, which has its polar expression in *KAPAY*, Purusha and Prakriti, the mirrors of 3 and 12.

Each sound or creative process mirrors others to interweave all creation together in virtually innumerable mantras or superstrings of sound. This is reflected in quantum physics, where it is seen that light reflects off the spiralling waves of creation to create the appearance of movement.

In Tantra, syllables can be transposed and united with other syllables in different combinations to create doorways into the universal creative process through "magic" number squares and geometric alignments.

This weaving of the cosmic tapestry can be used in meditation, ceremony and practical solutions in our lives, and much more.

May this book inspire, illuminate and guide you into alignment with the Blueprint of Creation.

Namaste!
Padma Aon Prakasha

Notes

1. Fritjof Capra, *The Tao of Physics*.
2. In the Cidambara Mahatmya 13:58-9, Apasmara, "looked like the agglomeration of all the devotees' sins. Shiva stood with one foot on Apasmara to destroy the sins they had accumulated in life after life."
3. *Kunchitangristava* 268.
4. The more enlightened a person is, the younger their astral or Ka body form, not necessarily physical body, will look like. This echoes Christ's saying, "Be ye like a child to enter the Kingdom of heaven."
5. *Srimad-Bhagavatam* 3:12:4.
6. Part of the *pratyabhijñā* philosophy of Kāshmīri Saivism, the earliest recorded exposition of non-dual Saivism, aside from the Saiva Agamas.
7. The Singularity of all forces of space-time that leads into the Void.
8. *The Sounds of Shiva* by Peter Harrison.
9. In the other telling of the story, one's ignorance is destroyed in the Tandava, and is then resurrected through the Lasya Dance, the softer more feminine movements of Shiva's Shakti.
10. The Maheshwara Sutra makes no mention of the Lasya dance of Ananda Thandavam. Either this is something different OR it is shared fully in the Maheshwara Sutra.
11. The Bible echoes this, where "God moved upon the face of the living waters" in order to create the universe.
12. The results were detailed in *Physical Review Letters*, 31 December 2004.
13. For a full explanation see Harrison's *Dhatu-patha: The Roots of Language*.
14. Ref: American Hindu University.

15. The Shakti Twins, a musical group based in the USA.
16. When sounding Sanskrit, sound the syllables deeply, precisely, with a soulful resonance in your voice and the part of the mouth, larynx and chakra it is resonating. Be aware and slow with it – not fast. There is no hurry in the Self or in its "languaging".
17. For example, there are 99 words for all the different varieties of the word "love" in Sanskrit.
18. Shiva reveals all ignorance by showing us all the elements that create us, as well as showing that creation is part of TatPurusha, the Universal pervading soul sustaining all sentient beings.
19. Rupert Spira, from an online talk, 14/05/20.
20. *The Hidden Meaning of the Lotus Sutra* (Hsuan-I, Gengi).
21. Zeki, S., Romaya, JP, Benincasa, DMT, & Atiyah, MF (2014) The experience of mathematical beauty and its neural correlates. *Frontiers in Human Neuroscience.*
22. The Four Modes of Sound are explained in the chapter on The Four Modes of Sound.
23. Bhartrhari, from the *Vakyapadiya.*
24. Panini's interpretation of this verse laid the foundation for Sanskrit grammar in multiple ways which one can investigate by reading his texts.
25. Walter Russell.
26. For example, the Backus Normal Form. Panini's notation is equivalent in its power to that of *Backus,* and has many similar properties. More about Sanskrit and Mathematics is shared in Appendix 1 at the end of the book.
27. *The Sounds of Shiva* by Peter Harrison.
28. Unless the grammar inputting code allows full expression from the moment as a child we start to be taught at school, which is not being done in the languages we currently use, i.e. English.
29. Sri Aurobindo, Elementary Roots of Language.

30. Joseph Rael, from *Being & Vibration*.
31. A Japanese sage gives the explanation that when people fall off buildings they shout AHH because they naturally wish to ascend. John Michell, *Euphonics*.
32. O is the end of all knowledge – Advaita Vedanta. In 0, there is nothing to know as everything is known. Anything multiplied by zero becomes zero.
33. *The Ancient Secret of the Flower of Life* by Drunvalo Melchizedek.
34. *Rk Veda*.
35. Bear in mind that all of creation is made up of only two shapes, the cube and the circle, masculine and feminine. The 8-celled cubic geometry is known as the Seed of Life, and stores the history of Creation of both the human being and the universe in this non-dual quantum pattern. This 8-celled cubic geometry holds keys to biology and technology in the quartz crystal cube, which stores data and could potentially make quantum machines in the near future.
36. Patanjali's Yoga Sutras 1, v. 24-26.
37. *The Sounds of Shiva* by Peter Harrison.
38. Patanjali's Yoga Sutras 1, v. 24-26.
39. Abhinavagupta's *Tantrasara*: The essence of the Tantras, Chapter Three: *Illumination of the divine method* (śāmbhava-upāya). Translated by Christopher Wallis. hareesh.org/blog/2019/1/5/the-divine-method-tantrasaara-chapter-three
40. Rupert Spira.
41. Sadhguru, from the book *Adi Yogi*.
42. Translating this information into human language may take years, but the person who has received it has transformed in some way by entering this mode of vibration.
43. Soviet Parapsychologist Inyushin.
44. Lawrence Blair *(13:38)*.

45. S. Buxton, *Darkness Visible.*

46. The Four Modes are mentioned at the beginning of creation after A I U, indicating that it is here that they are initially formed. I have placed verses 5 and 6 after the explanation of A I Un for the reader's clarity, yet it is important to know that these two verses about the four modes of sound spring forth from A I Un.

47. Bk 5, section 2.

48. Expanding on this, Lynnclaire Dennis of Mereon has shown that all sounds and frequencies arise out of this torus spiral, when viewed at different angles.

49. From his lecture "Opening the Doors of Creativity".

50. Of which more will be shared in the chapter on the fourth mode of sound.

51. From *Kashmir Shaivism* by Swami Lakshmanjoo. Note that this references the long and short letters of LRI and RI, and not just the singular letters of LRI and RI found in RLRK. This use of both the short and long letters of RI and LRI defines the process more clearly and in more detail. In the Maheshwara Sutra, RI is the intention *and* confirmation of Shiva returning to its own nature, and LRI is the establishing *and* the carrying out of the movements to come to rest in Its own nature.

52. *Tattvavimarsini* by Upamanyu.

53. Alexandra David-Neel and Lama Yongden, *The Secret Oral Teachings in Tibetan Buddhist Sects.*

54. David-Neel & Llama Yongden, *The Secret Oral Teachings in Tibetan Buddhist Sects.*

55. Dr Stuart Hameroff, Professor Emeritus, University of Arizona.

56. Lorin Roche, *The Radiance Sutras.*

57. Rupert Spira.

58. *Kashmir Shaivism* by Swami Lakshmanjoo.

59. Rusiko Bourtchouladze, *Memories Are Made of This.*

60. There are up to seven layers of meaning held within each word sound world, that connect through the *vrittis* to create subtle forms.

61. Rupert Spira.

62. *Tilapa Kriya.* It is the goal of many meditational practices to still the mind and Realize the unchanging Witness.

63. Dr Joe Dispenza.

64. This direct perceiving of reality is found in Pasyanti, one of the four modes of sound.

65. Nature, the world, the body-mind and the experiences it offers provide us with reflections and experiences, blissful and painful, that are played over and over again as we mistakenly identify ourself with this activity and forget about its source. This goes on and on until we remember the awareness behind this game. This happens through the witness.

66. David Hoffmeister.

67. What we observe is conditioned by our own method of interpreting and questioning. For example, when you are inside a building, a construct, you cannot see what it looks like and its dimensions. You are in it. Only when you are outside the building or construct can you see what it really is. The observer influences all that it sees, modifying the outcome of what is being observed, bringing forth patterns, programs and thought forms in this process.

68. Sri Nisargadatta Maharaj, from the book *I Am That.*

69. When this is seen and experienced clearly, Maya transforms into *svatantriya shakti* – pure Will energy.

70. This is different to the previous RLRK, as now there is something, rather than just potential.

71. Gnostic Apostle Thomas, Chapter 24.

72. Non-duality teacher Rupert Spira.

73. The Cosmic Egg is mentioned throughout the texts of the Vedic tradition as well as the Egyptian, Gnostic and many other traditions around the world.

74. The Theosophical Glossary.

75. In Vedic texts, Visnu is seen as emanating trillions of these eggs as the source of Hiranyagarbha, which then gives rise to Brahma as the Creator "hatching" this egg into manifest creation, as seen in the next chapter.

76. This can be experienced and played out through Tantra.

77. Herakleitos.

78. These interconnecting subtle movements are the blueprints and precursors to the toroidal spiral, which constantly interchanges between its centre point and its periphery; in effect, here and there at the same time.

79. Rabbi Stan Tenen. In the beginning of something also lies its end, as every question asked also contains its answer. As Christ said, "I Am the Alpha and Omega, the first and the last, the beginning and the end."

80. In the *Kubjikāmata-Tantra*, *Vijñānā* is the third (as *Eon* is the third) of the eight *Mātṛs* or measures of creation (Maya is a measurer of creation) born from the body of *Ātmī* or the Creator. All the different categories of souls are formed in this stage of creation, including the gods, as well as the karmas by which all souls are bound.

81. *The Crystal and the Way of Light: Sutra, Tantra and Dzogchen*, Chogyal Namkhai Norbu.

82. This is corroborated in the Sutra, as the big bang AIAUch occurs before the creation of space element. Space element is created in the next class of sound HAYAWARAt.

83. AU is the most powerful of all the Kriya Shakti (energy of action) sounds. It is said that Shiva is most vividly experienced in creation in this sound.

84. Walter Russell, *The Universal One*, pp. 182-183. Wave diagram from *The Ancient Secret of the Flower of Life* by D. Melchizedek.

85. Jain 108.

86. Nimbin Apothecary.

87. Blood flow through living vessels is more like a tornado than anything else: such a vacuum is necessary for producing a vortex. Blood cells spin on their own individual axes of rotation, being smaller spinning cells in a larger spinning vortex.
88. Walter Russell, *The Universal One*, pp. 182-183.
89. The AIM *bija* is the second most common *bija* after AUM.
90. Quantum gravity or the gravitational singularity developed during the Planck Epoch, up to 10-43 seconds after the birth of the universe in *AI AU*.
91. Foster Gamble, *Thrive* documentary.
92. The whole cycle of the Precession of Equinoxes over 25,920 years is known as a Year of Brahma.
93. In India, these true names are held in the Vedas, which hold the laws, sounds and regulations of creation.
94. Peter Johnson, "The 10 stages".
95. In human terms this could also lead to the most pride, i.e. "Look at me! I have created the very basis of everything!" When in fact, it is just one of the many phases of creation.
96. *Dr. John Hagelin, Science of Yagya: Global Transformation through Vedic Recitation.*
97. Vyoma marks a change of order from what has come before it in order to manifest the first subtle substance of matter as a medium for physicality.
98. Peter Harrison from *The Sounds of Shiva*.
99. Peter Harrison, *The Sounds of Shiva*.
100. Shankaracharya Santananda Sarasvati.
101. *Nature Physics Journal*, February 2019-03-02.
102. In the *Purushasûkta* of the *Rk Veda*, Vayu emerges from the breath of *Purusha*, the cosmic Man from whom all creation is birthed.
103. This is discussed in detail in the chapter on the Pranic Mind.

104. For the purposes of transformation, air, fire and water are the three easiest elements to work with, as they are all malleable and easy to get in touch with.

105. Dr John Reid.

106. Jeffrey Thompson, Center for Neuroacoustic Research, California Institute for Human Science.

107. Water interacts with nearly every part of the DNA's double helix.

108. Hexagonally in association with cellular protein structures.

109. With water's ability to absorb gas, it brings together life on the solid earth with those in the deep oceans.

110. We can see these movements through observation, or with the help of vibratory technologies, such as the CymaScope and Dr Emoto's work with water.

111. *Sensitive Chaos* by Theodor Schwenk, Steiner Press.

112. Emilie Conrad.

113. We can see these movements through acute observation, or with the help of vibratory technologies, such as the CymaScope and Dr Emoto's work with water.

114. Anne Stanford, CEO Braintree.

115. Cherionna Menzam-Sills.

116. Vimalananda, *Aghora III: The Law of Karma*.

117. Based on the *Rk Veda* and the 108 Names of Agni found therein, and the Yajur Veda. Thank you to Peter Harrison for compiling the 108 Names of Agni from the *Rk Veda*.

118. Maurice Bloomfield, *The Sacred Books of the East*, Vol. 42, 1897.

119. CBS News, 2017.

120. Thank you to Sri Yukteswar and Drunvalo Melchizedek for this meditation.

121. Maitri Upanishad.

122. Mahābhāṣya of Patañjali on the Sūtras of Pāṇini – Dr Kielhorn's edition – IV:2.36.

123. Vāj. Prāt. Adhāya 5.

124. In these streams of vibration are held the potential for the five sense organs to operate, as each sense organ connects to each of the tanmatras.
125. Swami Lakshmanjoo, *Self-Realization in Kashmir Shaivism*, by John Hughes.
126. Swami Lakshmanjoo, *Self-Realization in Kashmir Shaivism*, by John Hughes.
127. Claircognisance, or clear knowing, is also part of space element's gift, meaning you suddenly know something truly and clearly that previously you had no idea about.
128. From Krishnamurti Series II – Chapter 34 – "Listening".
129. *As seen in Ayurveda and the next class of sound Jha Bhan.*
130. AN Schore, *Affect Dysregulation and Disorders of the Self.*
131. Emma Pavey, Langley (British Columbia) Mennonite Fellowship.
132. Rev. Kelly Isola.
133. Every form can be seen as fire *and* as light. This is seen in the first Vedic deity of creation Agni, the first light and the first fire. Both are intertwined, as fire leads into light and comes from light. Another meaning of this saying of Christ is that when your eye is single, one pointed, focused in the third eye, then the whole body becomes flooded with light in samadhic meditation. This occurs as the breath and heartbeat slows down, and the body becomes flooded with bliss-producing hormones that reveal more light in the body. Hence, if thine eye be single in the pineal, the whole body is flooded with light.
134. These effects are designed to open the heart by direct opposites. Similar to the point in a movie, where a hero is killed tragically, with strains of soaring, angelic music playing behind.
135. *The Sophia of Jesus Christ*, Nag Hammadi Codex III.
136. *Principles of Ayurvedic Medicine* by Dr Marc Halpern, California College Ayurveda.

137. Disorders in our ability to taste are due to an imbalance in water element.
138. https://www.scribd.com/document/377077533/The-8-Clair-Senses-a-visual-guide-to-the-spiritual-psychic-senses
139. http://www.bbc.com/future/story/20150803-what-our-perspiration-reveals-about-us
140. *Kashmir Shaivism* by Swami Lakshmanjoo.
141. *Kashmir Shaivism* by Swami Lakshmanjoo.
142. *Vak* is associated with Saraswati, Goddess of speech, the Word and music, as seen earlier in *AIAUch*. Speech is also associated with fire element, bringing forth colour *and forms*. *Viradrupam* also means mighty, heroic, strong, potent *in form*.
143. *Proceedings of the Royal Society B: Biological Sciences.* 16 March 2022. Article by Tessa Koumoundouros.
144. The study was published online 12 May 2021 in the journal *Scientific Reports.*
145. Mirror neurons are a self-learning mechanism to coordinate perception and actions. This relates to abstract concept formation, language development, understanding, acquisition and cognition. Mirror neurons are part of a larger network that completes cross modal, motor and perception information.
146. Mirror neurons are *Necessary* but not a sufficient enough cause for evolution by themselves: their emergence and further development is probably also gene based.
147. The Singularity of all forces of space-time that leads into the Void.
148. For a fascinating breakdown of how this sequence was first discovered in India, by whom, and how it was used in music and Sanskrit, visit: http://guruprasad.net/posts/fibonacci-number-series-originated-ancient-india/
149. *Sam Baron*, Associate Professor, Australian Catholic University, 24 May 2022. www.sciencealert.com/what-would-happen-if-someone-moved-at-twice-the-speed-of-light

150. All movement arises in our human experience from our inner determination or will to act, an inner volition. All movements at this point only happen because of *GHA*, the seed of this volition.
151. *The Sounds of Shiva*, Peter Harrison.
152. Brhadaranyaka Upanishad 2:4.11.
153. Swami Sivananda.
154. Verse 9, translation by Lorin Roche in *The Radiance Sutras*.
155. Verse 50, translation by Lorin Roche in *The Radiance Sutras*.
156. In translating this verse, it can be said that the sense organs are born from the middle letters of each of the groups of the vargā-s of the alphabet i.e., *ja, bha, ga, ḍa, da* are the middle letters of cha-group, pa-group, ka-group, ṭa-group, ta-group of letters, and that these letters are the originators of the sense organs.
157. From an oral story shared by Peter Harrison.
158. Caraka Sutrasthana X1.
159. *The Sounds of Shiva*, Peter Harrison.
160. *The Mail Online*.
161. Living in darkness is said to force the brain to secrete more hormones, particularly DMT from the pineal gland.
162. Smell and taste are both chemical senses, unlike the other three senses.
163. Maffei, Haley, & Fontanini, 2012; Roper, 2013.
164. Neuroenology: How the brain creates the taste of wine, by Gordon M. Shepherd in the magazine *Flavour*, Volume 4, Article number: 19 (2015).
165. *The World as Power*, Woodroffe, pp. 103-5.
166. *The Sounds of Shiva* by Peter Harrison. Additionally, the word Prana has seven levels of meaning, just as all Sanskrit words do. Its meanings include the life-force and enlivening life flow, the connector, creator, the fuel for the mind and body, the universal creative power and the active force of

nature, the amount of energy we have to live, thrive, create and work with.

167. All quotes from Randolph Stone, found in *Esoteric Anatomy*.

168. Taittiriya Upanishad 2:3.1.

169. Bruce Burger, *Esoteric Anatomy*.

170. In many ancient societies like the Essenes and Tantric societies worldwide, the cleansing of the anus is seen to be of paramount importance, through enemas, colonics, tantric exercises and lovemaking.

171. *Udana* works with *apana* to precipitate the birthing process and this continues in life as the rising of the *kundalini* energies of awakening.

172. Brhadaranyaka Upanishad 5:3.1.

173. Peter Harrison, *The Sounds of Shiva*.

174. Curiously, Samana generates the sound heard by stopping the ears. When you are about to leave your body, you no longer hear this sound.

175. *The Sounds of Shiva*, Peter Harrison.

176. All quotes from Randolph Stone, found in *Esoteric Anatomy*.

177. Ayur Veda: Caraka Sarira Sthana 1:18-19.

178. Almaas, 2004, p. 509.

179. *The Visionary Window*, Quest Books.

180. *Bhagavad Gita* 18, v. 30-32.

181. Nikhilananda, 1947.

182. The book *Adi Yogi* by Sadhguru.

183. The book *Adi Yogi* by Sadhguru.

184. Swami Santananda Saraswati.

185. Rupert Spira, from an online talk, 14/05/20.

186. When one has any qualification, then one is subject to change. If one has no qualification, one is free of all circumstances, people and environments.

187. In Biodynamic Craniosacral therapy as pioneered by Franklyn Sills.

188. It can be sensed 50 centimetres or so around the body, and the tidal motion within the tidal body is sensed as a stable rhythmic phenomenon of 100 seconds (50 seconds of inhalation and 50 seconds of exhalation).

189. Private communications between Sills and Levine, 1999.

190. Franklyn Sills.

191. This then allows unresolved inert forces within us to shift into the quantum blueprint of the One.

192. Rupert Spira.

193. In the Egyptian: Amun Amaunet – *AH* and *I* Shiva-Sakti. Heh Hehuit – Maya-Ishwara *EOn*. Kek Kekuit – *KA Prakritim* and *PAy Purusam*. Nun Naunet – Brahma-Saraswati *AI AU*.

194. Hymn XC. Puruṣa, *Rigveda* by Ralph TH Griffiths.

195. Isha Upanishad 6-7.

196. Dr David Frawley.

197. *Biocentrism: How Life and Consciousness are the Keys to Understanding the True Nature of the Universe* by Robert Lanza and Bob Berman.

198. Swami Lakshmanjoo in *Self-Realization in Kashmir Shaivism*, by John Hughes.

199. Mircea Eliade.

200. This is seen in the black hole-white hole phenomenon later in the book.

201. Published in *Nature Materials*. Original article by David Nield, *sciencealert.com*, April 19, 2022.

202. Rupert Spira.

203. Sometimes this can be done in *nirbija samadhi* meditation, which burns up all these samskaric seeds of suffering spontaneously and totally.

204. Nicolya Christi, *2012: A Clarion Call*.

205. Jill Purce, *The Mystic Spiral*.

206. Peter Harrison, *The Sounds of Shiva*.

207. However, IF we use sattva to guide our actions, and tamas to ground us, this can all come into balance.

208. CS III-8:110.
209. This is hinted at in the Sutra when Nandikeshvara uses the word *Prabhu*, which is associated with one of the 108 names of Visnu; it is highly unlikely he would have used that word without a deeper meaning to it, such were the Rishis.
210. Bruce Burger, *Esoteric Anatomy*, NAB. "We can balance Tamas by working with our heels, sacrum and occiput through cranio-sacral work."
211. *Hindu Polytheism*, p. 27.
212. *Enlightenment! The Yoga Sutras of Patanjali*, translated by Maharishi Sadashiva Ishaya.
213. Robert Svoboda, Aghora 3.
214. The Gnostic Apostle Thomas, Chapter 24.
215. *The Sophia of Jesus Christ*, Nag Hammadi Codex III and the *Berlin Gnostic Codex* 84, 13-17.
216. This is similar to the Genesis Story in the Bible, when after six days God saw and declared that all that had been created was good.
217. We receive this Grace as the culmination of Self Realization. Vigrahah is also one of the 108 names of Shiva in *Shuddha vigrahah*, Lord of Pure Form.
218. Rupert Spira.
219. This is a very powerful experience in *samadhi* meditation, as one travels beyond the singularity point into no point in total bliss. One becomes Shiva.
220. *Kashmir Shaivism* by Swami Lakshmanjoo.
221. Shiva reveals all ignorance by showing us all the elements that create us, as well as showing that creation is part of TatPurusha, the Universal pervading soul sustaining all sentient beings.
222. In astrophysics, this is seen on a small scale to be a nucleus of a collapsing star which becomes intensely dense; this nucleus then forms a black "hole".

223. On a galactic scale, one of these voids could be the black hole at the centre of our Milky Way Galaxy.

224. On a cosmological level, physicist Itzhak Bentov found that in observing a quasar star ejecting matter from its nucleus, it turned back upon itself, forming an *ovoid egg shape*. (EOn) Matter follows this egg pattern from *A I Un* or a *white hole*, and as more complex forms develop they move around the egg, irresistibly drawn towards their polar complement destination of the black hole *HAL*.

225. Black holes can be detected by listening for the sounds they create in energy carried by gravitational waves. NASA's Chandra X-ray Observatory detected these sound waves, recording the deepest note ever detected from any object in the universe: 57 octaves lower than middle C, a frequency over a million-billion times deeper than the limits of human hearing.

226. Interestingly, physicist Nassim Haramein has shared that there is a black hole at the heart of each of our atoms, meaning we too are mini black holes!

227. Osiris' name in Egyptian is Asar or Usir.

228. § 1280a-b, Coffin Texts TS 777, CT VI, 410a-c.

229. § 616d-f, § 825, § 1629, § 828.

230. The Pyramid Texts talk about the nursing and feeding of the newborn as well.

231. Horus, born at the time of his father's resurrection, would not literally collect the parts of his father's body strewn across an entire country *unless he knew a way of reconstituting them into life*. Horus' inheritance was the blueprint of creation Osiris was holding, which is why he wanted to find him and resurrect him.

232. Shambhala, 1985, 341.

233. Interestingly, it is impossible for an observer to see a rainbow at any angle other than at 42 degrees from the direction opposite the light source. Even if an observer sees

another observer who seems "under" or "at the end of" a rainbow, the second observer will see a different rainbow – farther off – at the same angle as seen by the first observer.

234. In addition to this, Clement of Alexandria states that the Egyptian temple library was divided into 42 "absolutely necessary" books. Thirty-six contain the entire philosophy of the Egyptians memorised by the priests. The remaining six were learned by the *Pastophori*, or image-bearers. These 42 may be the 42 Books of Thoth, the Egyptian "god" of wisdom and science, which contained the knowledge of all creation, according to the 3rd century BC Egyptian priest-historian Manetho.

235. The more enlightened a person is, the younger their astral or Ka body form, not necessarily physical body, will look like. This echoes Christ's saying, "Be ye like a child to enter the Kingdom of heaven."

236. *Srimad-Bhagavatam* 3:12:4.

237. This comes from the trinity of shiva as *satchitananda*.

238. The Gospel of Matthew Lecture Five. https://rsarchive.org/Lectures/GA123/English/RSPC1946/19100905p01.html

239. Yeshua was of course a Hebrew in land, culture, upbringing. He studied with Rabbis, and of course used the language.

240. http://kabbalahsecrets.com/part-viii-phi-%cf%95-and-pi-and-the-42-letter-name/

241. In addition, there are 42 *Stations of the Exodus*, the locations visited by the *Israelites* following their *exodus from Egypt*, as recorded in Bible, Numbers 33.

242. Walter Russell.

243. One can be initiated into *Sivohum* from a living master or directly from Siva.

244. Resistance can be emotional wounds, traumas, physical issues, mental rigidity and "spiritual" beliefs.

245. It is said that it was originally established in Atlantis before Egypt.

246. In Hindu ceremony, the first six patterns of creation or classes of sound establish the "destiny pattern" or life path of an individual. This pattern is manifested and embodied in the foetus through earth element LAN.
247. From the Jain108 website. Thank you, Jain!
248. The geometry of the six circles around 1 is the geometry of the carbon 7 atom: light and strong.
249. The developers of this technology were the original creators of the iconic pyramid stage and sound system at Glastonbury Music Festival.
250. Da Vinci credits 1st century BC Roman architect Marcus Vitruvius Pollio as the first exponent of this idea, in his 13-30 BC book, *Des Architectura* in the chapter "Temples and the Human Body".

Appendix 1

Numbers

A change in language can transform our appreciation of the cosmos.

The dynamic relationships between micro and macro, subatomic and molecular, seen and unseen, can be described and expressed through mathematics. Maths is a relationship language *and* an object language, a language in which meaning derives from dynamic, ever-shifting, multiplying and dividing relationships as well as the individual properties of numbers.

Numbers and equations are an efficient and universal language, just like music and sound, to express evolving states in resonant structures of process and action operating in interconnected relationship. Mathematical language expresses how individual particles work together in waves and fields, as numbers are vibrations, with field, wave, and particle all expressed.

Numbers combine in context and interrelationship with each other, each number defining the next in its movement. Each number links to others in various processes in order to form new things.

All advanced civilisations use numbers as a unified platform understandable to all intelligences in the universe. Numbers reveal a unifying methodology that creates and is created by sound and geometry, all of which describe movements in relationship to each other in the holographic universe.

Holographic languages work in processes of movement and verbs, rather than static objects and nouns. Thus, holographic languages use terms such as "being" not "been", a perpetual "doing" not "done", always happening, always moving, for the universe is ever expanding. Life and language are not fixed, but are in interrelated constant movement.

This is true for certain aspects of maths and geometry as well. For example, Euclidean Geometry is a perfect language for describing and manipulating objects on flat surfaces, but is insufficient for describing curved space-time: for that a different language in non-Euclidean Geometry was needed, as Einstein discovered.

Mathematical languages are verby, with processes and transformation, so they can be used for quantum concepts. Maths is the language of quantum physicists, who have to rigorously prove their quantum theories using mathematical equations.

Sanskrit and Numbers

Sanskrit is a mathematical, quantum and holographic language. Each letter and sound has a number; many letters chain together to form a word, a sentence, an idea and an expression of creation. Ideas are thought-forms concretized through numbers as sounds, and sounds are the vibrations of all thoughts and processes in the Universe.

Numbers become codes in equations, which a spiritual scientist would translate as syllables forming a Sutra. Both are unifying methodologies that express and reveal perceptions about the nature of creation. With the combining of both, creation can be revealed in all of its facets.

Each Sanskrit letter has a number, with this system known as *Kaṭapayādi* (in the Kabbalah of Egypt and the Hebraic language this science of ascribing letters to numbers is called Gematria). The name *Kaṭapayādi* comes from the consonants *ka, ṭa, pa, ya*, which all signify 1. Thus the name *Kaṭapayādi* comes from these four consonants of 1.

Kaṭapayādi gives each of the 33 Sanskrit consonants numbers ranging from 1 to 9. The resulting letter-number associations are used to encode long strings of numbers into words. The resulting text is one that has a double meaning: a verbal and numerical one.

This system can be used scientifically to explore the nature of creation and decode the 42 letters of the Sutra, as well as the original Sanskrit commentary by Rishi Nandikeshvara. Undoubtedly there are some fascinating associations here.

Kaṭapayādi has traditionally been used to incorporate large numbers into text so they can be more easily memorised, and was most popular in the astral sciences in South India. The details of this system are expounded in Śaṅkaravarman's *Sadratnamāla*:

"*nañāvacaśca śūnyāni saṃkhyāḥ kaṭapayādayaḥ
miśre tūpāntyahal saṃkhyā na ca cintyo halasvaraḥ.*"
[The letters,] na, ña and the stand-alone vowels such as A represent zero. The numbers 1 to 9 are represented by the consonant-groups beginning with ka, ṭa, pa, and ya [going serially along the alphabet series]. In a mixed/conjunct consonant, only the last of the consonant counts. A consonant with a vowel-sound is not regarded in this enumeration.

The correspondence between Sanskrit numbers and letters is:

1	2	3	4	5	6	7	8	9	0
ka	kha	ga	gha	ṅa	ca	cha	ja	jha	ña
ṭa	ṭha	ḍa	ḍha	ṇa	ta	tha	da	dha	na
pa	pha	ba	bha	ma	-	-	-	-	-
ya	ra	la	va	śa	ṣa	sa	ha	-	-

To decode the Maheshwara Sutra sounds numerically, and then infer what they mean, would unfold a wealth of information about the nature of creation. For example, the vowel sounds are all given the number 0. So, the first class of sound about the beginning of creation and quantum field *AIUn* is 0, 0, 0.

The 33 consonants that have numbers are understood to come from the Feminine *Code of Prakriti* – the energy information for matter existence. An interesting correlation to this is KAPAY Purusha-Prakriti, the formless light and material nature, which are the numbers 1, 1.

In decoding the numbers of the Sutra, we arrive at:

AIUn = 0, 0, 0
RLRK = 2, 3, 2
EOn = 0, 0
AIAUch = 0, 0, 7
HAYAVARAT = 8, 1, 4, 2
LAN = 3
NAMANNANAM = 5, 5, 5, 5, 5
JHABHAN = 9, 4
GHADHADHASH = 4, 4, 4
JABAGADADASH = 8, 3, 3, 9, 9
KHPHCHTHTHCATATAV = 2, 2, 7, 7, 7, 6, 6, 6
KAPAY = 1, 1
SHASASAR = 5, 6, 7
HAL = 8

Appendix 2

Technical Note

This book is not written or designed to be a grammar primer or linguistic analysis. I leave this to far more qualified people than me. This book is an experiential decoding of the creative process, designed to be shared as a practically applicable and eminently usable transmission.

The Maheshwara Sutra contains the fundamental sounds of the Sanskrit Alphabet. The long form of the first 5 vowels are not overtly included, nor are the special sounds *ksha, jna* and *tra* because they are combined or extension sounds.

The consonant endings at the end of each of the 14 Classes of Sound create forms. These consonant endings point to the first letter of each Class of Sound as being a key. The consonant endings create a code to put together groups of sounds. For example, *ac* represents all the vowels and *yan = y,v,r,l.*

Acknowledgements

Thank you to my sister and godmother of my son, Aina, for looking after us both as I took the tens of thousands of hours necessary to write this book.

Thank you to my mother Meera for sitting with me as I hounded her many times about certain verses in the Sutra! Having passionately learnt Sanskrit in India since she was eight years old, her authentic take on some of the verses and her passing on of oral tradition stories and wisdom was an invaluable grounding and clarifying aid in translating parts of the Sutra.

Honour to the *Shiva Jyotir Lingam* of Kedarnath, where this book began and where I was given my name.

Jai Maheshwarani! Whose wisdom is the Maheshwara Sutra, who is the entire universe, and without whom nothing would exist.

About the Author

Padma Aon Prakasha is an initiated *Saivite* named by the Head Priest of Kedarnath, the *Jyotir Shiva Lingam* in the Himalayas. His lineage is Kashmir Saivite traced back to the 1700s, where his ancestors were Saivite Pandits and Swamis in Kashmir. The *ishta devata* or personal god form of his ancestors has always been Shiva, and his *gotra* or bloodline is descended from Rishi Vasistha.

Padma's spiritual training began at age four with his formal initiation into the Brahmin Lineage, which led to his reading the *Bhagavad Gita* and many of the European philosophers by age seven. At age 12 he realized he had read enough, and decided to viscerally experience the consciousness of what he had been reading. Playing music seemed the best way to do this!

At age 20 an experience of Unity Consciousness changed his life forever. Shortly thereafter, Padma had a Near Death Experience Initiation in one of the four most sacred sites in India: Badrinath. Further initiations into the *Rk Veda* through the Arunachala Sampradaya led to Padma's most powerful Initiations through an awakened teacher called Sri Om in London, after which he entered into *nirvikalpa samadhi*, Samadhi with seed, for a period of two months.

Padma's work has touched millions through his books, teachings, transmissions, and the many teachers he has taught. Inspiring, moving, direct and a breath of fresh air, Padma has been teaching worldwide since 1997 as a potent evolutionary catalyst, bringing together ancient and modern, esoteric and practical, to create change for those who truly desire it.

Padma's books reveal different aspects of the full potential of the soul. His books include: *Dimensions of Love* (O-Books), *The Power of Shakti*, *Womb Wisdom* and *Sacred Relationships* (Inner Traditions), *The Christ Blueprint* and *The Nine Eyes of*

Light: Ascension Keys from Egypt (NAB/Random House). Since 1997 he has taught workshops and retreats in over 20 countries worldwide.

Padma is a master of vibrational medicine through sound, and translates the art and science of vibration to create immersions that are moving and alchemical. A globally distributed music producer, Padma has performed worldwide, producing two albums for Sub Rosa/BMG: *Rhythmic Intelligence*, and *Song of Light*.

His album *Life Cycles* accompanies the book *Womb Wisdom*, voted one of the top 30 spiritual books you must read. His latest album *The Soul's Birth* merges the science of psychoacoustic frequencies with Sanskrit and the latest in sonic technology to create a re-birthing soundtrack for the soul.

His TV documentary work involved co-discovering the first-ever-found mummified Tibetan Lama for *Mystery of the Tibetan Mummy* (Discovery Channel) and narrating Emmy and Peabody Award-winner Stephen Olsson's Cannes and Sundance Festival Documentary *Sounds of the Soul* about the Fez Festival of World Sacred Music.

As a multimedia producer he has created original music and visual content based on the sacred sites of the world for the Moscow Youth Olympics. He has also produced animated soundscapes and acted as a Choir Director for live performances, training choirs to sing multiple interlocking Names of God in five sacred languages, which resulted in the CD and live performance of *The ONE*, which played in theatres and festivals in UK, USA, Australia and New Zealand.

Padma has supported the environment by leading groups on curated Transformational Journeys to sacred sites in 15 countries on 5 different continents for the last 20 years. He has appeared on or in BBC Radio 1, *Times of India*, William Henry's Dreamland Radio, H_2O Radio Network, *XLR8R*, Straight No Chaser, *Variety* magazine, Watkins' *Mind Body Spirit* magazine, *Spirit & Destiny*, *Kindred Spirit* and *Christ Consciousness* magazines, amongst others.

Bibliography

Sri Nandikesakasika by Rishi Nandikeshvara
Tattvavimarsini by Upamanyu
The Sounds of Shiva by Peter Harrison
Adi Yogi by Sadhguru
Kashmir Shaivism by Swami Lakshmanjoo
Self-Realization in Kashmir Shaivism: Oral Teachings of Swami Lakshmanjoo by John Hughes
Investigating Reality by Amit Goswami
The Nine Eyes of Light by Padma Aon Prakasha
The Science of Sound by Padma Aon Prakasha
The Radiance Sutras by Lorin Roche
Esoteric Anatomy by Bruce Burger
Sensitive Chaos by Theodor Schwenk
The Ancient Secret of the Flower of Life by Drunvalo Melchizedek
Rk Veda, Bhagavad Gita, Patanjali's Yoga Sutras: Various translations

Videos
Thank you to Dr Vasant Lad
Dr Robert Svoboda
Rupert Spira

Sharing This Book
This book took tens of thousands of hours to write over a 17-year period in five different countries. It would be in integrity for you, the reader, to not share this Book to others for free. Instead, share a link with other people to purchase the book and support this Work. Thank you.

Other Titles by this Author

The Power of Shakti: 18 Pathways to Ignite the Energy of the
Divine Woman
978-1594773167

The Christ Blueprint: 13 Keys to Christ Consciousness
978-1556438844

Womb Wisdom: Awakening the Creative and Forgotten
Powers of the Feminine
978-1594773785

The Nine Eyes of Light: Ascension Keys from Egypt
978-1556438905

Sacred Wounds: Original Innocence
ASIN: B005MGOZEE

Dimensions of Love: 7 Steps to God
978-1780995137

Sacred Relationships: The Practice of Intimate Erotic Love
978-1620555491

BOOKS

SPIRITUALITY

O is a symbol of the world, of oneness and unity; this eye represents knowledge and insight. We publish titles on general spirituality and living a spiritual life. We aim to inform and help you on your own journey in this life.

If you have enjoyed this book, why not tell other readers by posting a review on your preferred book site?

Recent bestsellers from O-Books are:

Heart of Tantric Sex
Diana Richardson
Revealing Eastern secrets of deep love and intimacy to Western couples.
Paperback: 978-1-90381-637-0 ebook: 978-1-84694-637-0

Crystal Prescriptions
The A-Z guide to over 1,200 symptoms and their healing crystals
Judy Hall
The first in the popular series of six books, this handy little guide is packed as tight as a pill-bottle with crystal remedies for ailments.
Paperback: 978-1-90504-740-6 ebook: 978-1-84694-629-5

Your Simple Path
Find Happiness in every step
Ian Tucker
A guide to helping us reconnect with what is really important in our lives.
Paperback: 978-1-78279-349-6 ebook: 978-1-78279-348-9

365 Days of Wisdom
Daily Messages To Inspire You Through The Year
Dadi Janki
Daily messages which cool the mind, warm the heart and guide you along your journey.
Paperback: 978-1-84694-863-3 ebook: 978-1-84694-864-0

Body of Wisdom
Women's Spiritual Power and How it Serves
Hilary Hart
Bringing together the dreams and experiences of women across the world with today's most visionary spiritual teachers.
Paperback: 978-1-78099-696-7 ebook: 978-1-78099-695-0

Dying to Be Free
From Enforced Secrecy to Near Death to True Transformation
Hannah Robinson
After an unexpected accident and near-death experience, Hannah Robinson found herself radically transforming her life, while a remarkable new insight altered her relationship with her father, a practising Catholic priest.
Paperback: 978-1-78535-254-6 ebook: 978-1-78535-255-3

The Ecology of the Soul
A Manual of Peace, Power and Personal Growth for Real People
in the Real World
Aidan Walker
Balance your own inner Ecology of the Soul to regain your
natural state of peace, power and wellbeing.
Paperback: 978-1-78279-850-7 ebook: 978-1-78279-849-1

Not I, Not other than I
The Life and Teachings of Russel Williams
Steve Taylor, Russel Williams
The miraculous life and inspiring teachings of one of the World's
greatest living Sages.
Paperback: 978-1-78279-729-6 ebook: 978-1-78279-728-9

On the Other Side of Love
A woman's unconventional journey towards wisdom
Muriel Maufroy
When life has lost all meaning, what do you do?
Paperback: 978-1-78535-281-2 ebook: 978-1-78535-282-9

Practicing A Course In Miracles
A translation of the Workbook in plain language, with
mentor's notes
Elizabeth A. Cronkhite
The practical second and third volumes of The Plain-Language
A Course In Miracles.
Paperback: 978-1-84694-403-1 ebook: 978-1-78099-072-9

Quantum Bliss
The Quantum Mechanics of Happiness, Abundance, and Health
George S. Mentz
Quantum Bliss is the breakthrough summary of success and
spirituality secrets that customers have been waiting for.
Paperback: 978-1-78535-203-4 ebook: 978-1-78535-204-1

The Upside Down Mountain
Mags MacKean
A must-read for anyone weary of chasing success and happiness
– one woman's inspirational journey swapping the uphill slog for
the downhill slope.
Paperback: 978-1-78535-171-6 ebook: 978-1-78535-172-3

Your Personal Tuning Fork
The Endocrine System
Deborah Bates
Discover your body's health secret, the endocrine system, and
'twang' your way to sustainable health!
Paperback: 978-1-84694-503-8 ebook: 978-1-78099-697-4

Readers of ebooks can buy or view any of these bestsellers by
clicking on the live link in the title. Most titles are published
in paperback and as an ebook. Paperbacks are available in
traditional bookshops. Both print and ebook formats are
available online.
Find more titles and sign up to our readers' newsletter at
http://www.johnhuntpublishing.com/mind-body-spirit
Follow us on Facebook at https://www.facebook.com/OBooks/
and Twitter at https://twitter.com/obooks